Something There Is

Seeking a Rational Faith for Our Children

Something There Is

Seeking a Rational Faith for Our Children

David Sayre

PETER E. RANDALL PUBLISHER
Portsmouth, New Hampshire
2014

Printed Edition ISBN: 978-1-931807-82-1

Library of Congress Control Number: 2013923283

Published by
Peter E. Randall Publisher
Box 4726
Portsmouth, NH 03801
www.perpublisher.com

Book design:
Grace Peirce

Something there is that doesn't love a wall,
That sends the frozen-ground-swell under it,
And spills the upper boulders in the sun;
And makes gaps even two can pass abreast.
                    —Robert Frost, "Mending Wall"

Preface

THIS IS THE STORY of a search for a rational faith. It is really many stories, recorded over many years in places of great strivings—for comfort or meaning, for peace or power, connection or freedom, for understanding, healing, forgiveness, identity, worth, beauty, courage, for a way to live, for love. Some of the strivings seem noble and some base; but they are all real, as the places and people are real.

All the events in this story are factual, with only their sequence altered. A few of the names of people and places, all of which are also true, have been changed to avoid intrusion. All such cases have been marked with a footnote.

I tell the story as participant and observer, but not as a central figure. Its validity, and I hope its interest, derives from those encountered in the places from which I write. It is told in a series of letters to our children and their children, which I hope to be an engaging and honest style of writing.

Many of us today are unable to accept religious dogma without a rational base. Yet in these encounters over the years I hear a persistent longing and a kind of faith—that is, a commitment and extension and trust—in truth and in each other. These encounters lead me to believe that a rational faith can be found and that the things we hold sacred are real, after all.

Yes, I understand your question, and no,
I don't have any easy answers.

But I have some ideas about where to look.

Contents

Preface	vii
Contents	x
Emerson Hospital*	1
Harvard, Massachusetts	7
The Arlington Street Church, Boston	13
Radio Communications Corporation[N]	25
The Development Center[N]	41
Middletown Prison[N]*	35
The Pentagon, Washington, DC	51
The Newfern School for the Retarded[N]*	57
The Fitness Club[N]	63
La Musée Marmottan	77
Shaftsbury, Vermont	91
Monhegan Island	99
The Metropolitan State Hospital*	107
Dorchester, Massachusetts	117
Hawikkuh, Zuni Land	125
Manhattan Island	139
Hard Scrabble Wash	145
The Massachusetts Institute of Technology	157

---

* Four of these letters were written from institutions that changed substantially, in population, treatment, and structure, between the time of the author's experience in them and the letters' publication. This does not alter the events, experience, or lessons of that time.

Gloucester Cathedral 175

The Berkshire Hills 193

Tilden Pond 207

The Board Room 217

The South Bronx 227

Northfield Mountain 237

The Emergency Room 253

Harvard, Massachusetts 273

Glossary 284

Index 291

Emerson Hospital

HERE NATURE TOLERATES improbabilities sustained against her downward slope. Here our local evidence of life begins, and here it ends, casting from its dawns and evenings equal but opposite shadows. Standing in this ambivalent shade we look out on a probabilistic universe that shows no preference except for equilibrium, a universe tumbling downward to disorder, charting no course except for random walks. Yet here and there the unlikeliest of things appear: a capacity to organize and communicate, a freedom to choose, an evidence of intelligence.

The hospital is full of shadows and full of light, falling in varying directions upon us as we come and go. One cannot say that life exists precisely here or there, in molecules or stars, or starts or ends, or can be held or lost. The great improbability eludes our snares; but if we cannot own it, perhaps we can share it. Perhaps, as we wait for you, we may begin to understand what it is.

Depending on your outcome, so to speak, your mom and I have selected a couple of proper names for you. I hope you will like them. In the meantime, I am addressing a snow-white mound on her belly, rather like the bumps that skiers call "moguls," and it occurs to me that your growing importance to us evokes the same name, so for a while I shall call you my little Mogul.

This morning I am writing to you from the hospital's Board Room, also full of moguls. We have just concluded a breakfast meeting of the Executive Committee, called to address three issues presumedly beyond the purview of Management. The first is of immediate and one might say organic interest to you: whether or not this medium-sized community hospital can sustain a full obstetrical service. On that issue a prominent member of our medical staff has just presented his committee's findings, providing the high point of the meeting thus far (breakfast excepted): "We find no professional impediment to maintaining a first-class OB/GYN service at Emerson. Our problem is neither marketing nor service quality, but the number of births in our catchment area. These towns simply do not produce enough babies. Ladies and gentlemen, this is not a medical problem; it is a lay problem." Well, Mo, your mom and I are doing our part.

Now this issue may seem of life-and-death importance to you, given your present undignified position, but our greatest challenge loomed in the next item on the agenda: how to handle that most perplexing, ambiguous, frustrating, complex problem ever to burden the wisest and most sophisticated hospital trustee, namely, Parking. Even the Chaplains lost their tempers. Somehow we muddled and compromised our way through, and the meeting turned to face the ghost of the future.

The Board Room is at the end of an upstairs hall in the old cottage overlooking the Sudbury River, given by the Emerson family early in this century to be a hospital for the community of Concord. This part of the cottage is unpretentious, more roughly finished than the elegant entrance and downstairs halls. Extending in several directions are newer wings, added one at a time to invite the evolving practice of what Lewis Thomas called The Youngest Science. [1]

To keep pace with that science, the hospital has become a Tin Woodsman, all new parts substituted for the original organic limbs, so that only the beginning intention remains. Among its viscera, a central boiler plant beats faithfully in all seasons, laboring even in summer because of the proliferation of simultaneous heating-and-cooling systems driven by inefficient controls. Over the years more and more offices have been fit into its larger spaces to accommodate the endless complexification of record keeping and reimbursement requirements, so that the administrative hive now threatens its sticky metastasis over the whole giving of medical care. It is this threat of inefficiency and bureaucracy, and the new specter of competition, that so exercise the medical staff representatives this morning.

Not far up the road, the LaMay Clinic^{N} * has begun digging a large hole next to our major highway. To the practitioners of medicine as we have known it, the clinic construction site is a gravitational black hole into which our innocent patients will be inescapably drawn with promises of high-tech, drop-in, team-managed treatment. Like our new Mental Health Center, the clinic will intrude with salaried physicians into the sanctified doctor-patient relationship, and impose financial considerations on medical judgments. The Board nominates me to a task force to visit the president of LaMay and try to negotiate a truce in the war for medical customers.

There will be truces, and more battles, and a growing frustration, born of the inescapable fact that our service is vastly different from health care a generation ago, and that its exploding capabilities are escalating costs and complexity beyond our control. Clinics and salaries and the new health maintenance organizations are only the beginning of issues that will tear apart the medical fabric we trusted. We all sense that a gap has developed between our ability to heal and our ability to administer and pay for the healing equitably, and that this gap is widening. I will serve on a dozen boards and our own family will have seven doctors of medicine and psychology in it, but I will not find evidence of the narrowing of that gap.

---

* The superscript ^{N} will be applied to any name changed, the first time it appears.

Board meetings adjourn by attrition. One by one the trustees remember other obligations of the day, and a diminishing group lingers to clean up the minutes and set another date. So I am free to go.

* * *

BUT SOME HERE cannot. I have a small friend downstairs who will not go home. The injustice of his suffering, and the anguish of his parents, are much on my mind as I write to you. We are so full of hope, and they of despair.

Downstairs from the old Board Room and into the newer wings, the hospital assumes its purpose. Parking and certificates of need and forms of reimbursement recede quickly. Here the hard, uncertain business of healing is worked, and here our veneers are quickly washed away by tears. Unconsciously I accelerate my pace, passing the living and the dying in our corridors, averting my glance from those I do not know to reach the little boy I do.

Matt[N] does not recognize his friend. He is carrying too much pain for a small body, and the burden has crushed him inward, beyond our touch. I see us all in that withdrawal from the world. The long reach that our minds can make in comfortable times draws tight around us when pinched by pain or fear or grief. There is plenty of that here.

Is that why Matt's suffering touches me so close? Am I seeing myself in that small crib, pinned down and shrunk and isolated by that which none of us escapes? Or am I identifying with his parents, my friends? What shall I say to them, Mo? My embrace, my words do not halt the disease that seems so unfair and somehow unnatural. They go home in turns, exhausted and despairing. How shall we comfort them, Mo? Where shall we look?

Something in the recognition of myself in Matt calls me, something about farther signals and the reach of our minds. I remember other friends engaged in listening for the very farthest signals, messages that might tell us more of who we are.

Notes

[1] Lewis Thomas, *The Youngest Science: Notes of a Medicine-Watcher* (New York: Viking Press, 1983.)

Harvard, Massachusetts

THE TOWN OF HARVARD is not to be confused with the university of the same name, although there are some connections, one of which I have come to explore. This Harvard is apple orchards and stone walls through the woods, evidencing the phases of its New England history: the great ice scattering New Hampshire granite south to Long Island; the primeval woods; then all farms, extracted at heavy effort from the woods and stony fields; then woods again.

One cannot visit this Harvard without Robert Frost. His verse springs spontaneously from the rough ground in all seasons, recalled in that gravelly voice so befitting the New England poet. For a moment in the morning chill, it seems I could listen intently enough to transcribe the earth-songs the way Frost did—an illusion decisively shattered when I try it.

Among the orchards, on the eastern slope of a hill not far from the little laboratory where I work, a modest sign marks the entrance to the Oak Ridge Observatory, Center for Astrophysics. This is property of Harvard University, the George R. Agassiz Station of the Harvard College Observatory, affiliated with the Smithsonian.

The station is smaller than its titles, and rather out of use. At the end of a path leading east is the strange telescope I have come to see. Its receiver and processor, where the serious work is done, are waiting in a small, dull structure to the rear, but its arresting feature is the antenna. Eighty-four feet across, the parabolic eye looks out on the universe, attending, patient. Whether we listen or not, no electromagnetic radiations of sufficient wavelength escape its notice. The shorter waves slip through its aluminum fingers, but they are not its job; this is a radio, not an optical, telescope, and it is here to catch radio-frequency signals of understandable purpose that may come our way. It is at rest this morning, facing south and only a little skyward, with its azimuth secured by a rusty padlock. A great stillness falls with the snowflakes on my notes.

* * *

ON ANOTHER SUNDAY morning, in September of 1985, Producer-Director Steven Spielberg came here with Professor Carl Sagan to activate the radio receiver he had donated to the search for extra-terrestrial intelligence. By then, Project Sentinel had been listening on 128,000 channels for two years without result. A "Mega-channel Extra-Terrestrial Assay" would immensely accelerate the search, allowing 8.4 million microwave channels [1] to scan 80% of the sky every 200 days. Funded by NASA and the Planetary Society (Carl Sagan, President), the META Project hoped to detect recognizable signals from great distances that would carry evidence of intelligent life elsewhere. Spielberg brought $100,000 and his infant son Max to the event, full of hope for the future. Mal Jones, an inventor and farmer from Martha's Vineyard, had painted the "symbol of unity" on the antenna's pedestal, its blue and yellow and white metaphor echoing the poem below it, concerning respect for our fragile blue home and the universe beyond. [2]

In the meantime, NASA was completing the R&D phase of an even more ambitious search. The Search for Intelligent Life (SETI) Project envisioned two complementary searches over the next several years, one using NASA's Deep Space Network of 34-meter radio telescopes and the other the 305-meter dish cut into the jungle near Arecibo, Puerto Rico, combined with other radio telescopes of 64- and 100-meter diameters. The first would be an all-sky search for "beacons," covering the 1-10 gHz frequency range and a few spot bands of high likelihood; the second a more targeted search aimed at Sun-like stars within 80 light-years, plus a few nearby galaxies and clusters. The first would listen for a second or so and move on, using a relatively wide-band receiver, whereas the second would listen to each target for minutes at a time with a much narrower-band receiver. Carl Sagan had in fact proposed such a search back in May of 1975. [3]

Perhaps these searches will be successful, perhaps not; as I write, no one knows. The suspense is more bearable than its resolution; our only comfortable condition on this question is ignorance. An unambiguous answer to the question of whether we are alone is hardly tolerable, even in prospect. Both the yes answer and the no answer carry enormous implications, which we can escape only so long as we remain unsure.

On this chilly morning the wind conspires with the antenna's skeleton, whispering Carl Sagan's question in endless reprise: are we alone? On the pedestal the painting and poem have faded. The dignitaries and reporters have left, the search moved on mostly to larger apertures and faster processors. But in the wind the haunting question lingers, and all that waits behind it. Are we alone?

I want to say that we are never alone. On mornings like this, Nature seems to sing of a prevailing truth, a universal potential of conscious life, a wider home. Can we hear such a song in rational tones?

There is, of course, the earliest hum: that primordial, vestigial radiation from the first escape of photons following the "big bang" of our universe's beginning. Arno Penzias and Robert Wilson, two radio engineers at Bell Labs, discovered that left-over echo by accident in 1965, "still quiring to the young-eyed cherubins" at about twenty-eight octaves above middle C, a rather low note as cosmic singing goes. [4][5]

But what we long for is a more personal song, a message from the universe that we are loved. We start by listening for the sounds of other life, which is what is going on at Harvard and SETI. So far no conclusive results can be reported. Perhaps the channel is wrong: why should a different life form communicate via modulation of electromagnetic radiation at radio frequencies? Our search, like our assumptions, is anthropomorphic.

I follow the antenna's gaze and try to listen beyond our sky. What do we hope to hear? How shall we know when we have heard it? What are the songs of intelligent life by which we shall know those who we hope are singing to us?

The search reflects our own questions about our identity. What sort of creatures are we, do we wish to be? How do we communicate, connect? Where in this universe is our home?

Perhaps someone is out there. Perhaps they are like us, but a little better, someone we would like to be. Perhaps finding them would help fulfill our instinct toward unity, help assuage the pain of giving up our old religious myths, help us with meaning and with grief. Perhaps they would help answer your question, Mo.

Notes

[1] Radio engineers apply the term "microwave" rather loosely to wavelengths somewhat shorter than our fingers, which were very short indeed early in radio days. Infrared, visible light, ultraviolet and x-ray radiation are also electromagnetic "waves," of progressively shorter wavelengths, now routinely used for purposes unimaginable when I was a boy with a vacuum-tube radio. The relation of wavelength to frequency is quite simple: their product equals the speed of light, about 300,000,000 (3×10^8) meters per second in a vacuum. Thus a 1mm wavelength signal would correspond to a frequency of 3×10^{11} cycles per second ("Hertz," abbreviated Hz.)

[2] Ann Levison, *Harvard Post*, Oct. 4, 1985. Jones's painting was based on a design by an Illinois farmer distressed by what he considered the colonialism of astronauts placing the U.S. flag on the moon. (Ten years later, on October 30, 1995, the signal processor at the Harvard observatory was upgraded to increase the speed of search and decrease false alarms, using private funds. By then, government funding of the more ambitious searches had been largely curtailed.)

[3] Carl Sagan and Frank Drake, "The Search for Extraterrestrial Intelligence," *Scientific American*, May 1975, 80–89.

[4] My musical analog is based on the following arithmetic: microwave background radiation (from the original escape of photons from the primordial soup of our universe's beginning) shows up with a characteristic spectrum that one would expect from a heat source at 2.73 degrees Kelvin. This is the temperature to which "left-over" radiation from that escape is predicted to have cooled as the universe expanded. A 3mm wavelength, around the center of the microwave band, corresponds to 10^{11} Hz. Middle C is often tuned to about 2^8

Hz; and each octave in Western music doubles the frequency (tone) of a note. Doubling 28 times raises this to 2^{36}, equal to $10^{10.84}$ ($2=10^{0.30101}$.) Thus if I raise middle C by 28 octaves I get to a middle frequency of the radio noise heard by the Bell Labs engineers, wherever they turned their radio telescope toward the sky.

[5] The fragment of Shakespeare appeals to me as a poetic equivalent of the vestigial hum of our universe's beginning. It is from *The Merchant of Venice*, Act V, Scene 1:

> Look how the floor of heaven
> Is thick inlaid with patines of bright gold:
> There's not the smallest orb which thou behold'st
> But in his motion like an angel sings,
> Still quiring to the young-eyed cherubins;
> Such harmony is in immortal souls;
> But whilst this muddy vesture of decay
> Doth grossly close it in, we cannot hear it.

The Arlington Street Church, Boston

I N THE BRIEF AFTERNOON of the nineteenth century, all the order on Boston's map was ransomed from the sea. The "Back Bay" of Boston rose from the Atlantic Ocean at the price of leveling hills nearby, as the press of commerce and comfort swelled the neck of the peninsula and defined the Charles River as it flows today. First disciplined was the Public Garden, arrayed westward from Charles Street, the original shoreline; then the alphabetical streets, Arlington, Berkeley, Clarendon, Dartmouth, Exeter...

I am on Arlington Street this morning, in the first substantial public structure erected in the Back Bay, the Arlington Street Church. Its Victorian builders trusted in their commerce and their future, in freedom and education, and in brownstone. That sedimentary rock, quarried along the Connecticut River, cuts nicely and looks fine with brick, but it does not like New England winters or car exhausts, and today the church shows its seasons.

Its builders' trust in education proved more dependable, although we have not yet got it quite right. These were early Unitarians, many of them Abolitionist leaders, their faith reinforced by what seemed the recent victory of the North's Right over the South's Wrong. That we have since discerned a wider purchase of Wrong does not detract from their dedication. Here William Ellery Channing cast a humanist perspective on Victorian religions and articulated a pursuit of truth above dogma. Here the Mays and the Eliots worshipped, families of farmers and preachers and underground railroaders who would become writers and scholars of great distinction, who would inter-marry and give Abigail Eliot to education and to me. [1]

A century has gone by, and I am here on a less spiritual assignment. The Arlington Street Church furloughs its choir in the summers and hires a quartet of soloists (at rather modest fees) to provide the anthems and responses appropriate to Unitarian worship. I have won an audition and am here to meet the other three survivors and preview the summer's music. This gives us an opportunity for the first time to rehearse in the loft and to enjoy the acoustics. (We try a Saint-Saens *Magnificat*, operatic and demanding, and as the last echoes fade the tenor turns to me and asks, "Have we stopped?")

* * *

OUTSIDE THE CITY, the history of Protestant churches is also organically linked to the history of towns: the Massachusetts legislature did not grant town charters in its early history, but parish charters. The state, and much of New England, is dotted with first parishes, now largely Unitarian-Universalist like the Arlington Street Church, but with their roots deeper and their trunks split by the Trinitarian-Unitarian schism. One of these parishes is in Lincoln, Massachusetts, where I go to sing tomorrow.

The simple structure across from Lincoln's town library recites the traditional (pre-Victorian) paradigm of New England houses of worship: white clapboards, symmetrical architecture, tall fenestration and steeple, center entrance—and the descendants of the original people of the parish, those who chartered the Town, still worshipping here, still singing the old hymns.

Tomorrow it is a Bach cantata, accompanied by a small orchestra. The instrumentalists and the other soloists are full-time musicians, and I am nervous. (Actually they are, too; one does not conquer that queasy feeling and accelerated heartbeat, but rather learns to anticipate and perform with it.) As we warm up and rehearse a few of the more intricate passages, I wonder at the skills that the instrumentalists have developed. They are the real musicians here, although they will get little of the credit. I try to focus on the privilege of performing great music with them and on communicating some of Bach's intention to the congregation.

Some of this music is almost painfully beautiful, arresting, even startling, and one can understand how people hear in it the voice of God. Bach's music in particular is orderly, intricately structured, mathematical; yet it is endlessly creative and its long, flowing phrases afford substantial interpretive opportunity in performance. It therefore communicates successfully: enough surprise to sustain the flow of information, enough structure to provide a grounding or context, enough elegance to convey a long message in a short space. I love to sing Bach, and most people love to hear it—well, in reasonable doses.

During rehearsals, when one has time to reflect, it can move us to tears. But I know better than to allow such emotions during a performance, because they do ugly things to the voices of humans. Besides, I have to concentrate through a long string of very black notes in proper synchronization with better-trained voices and instruments.

When it is over I wonder again at what it is that so moves us, why humans have sung so persistently through their whole history. There has been music almost as long as there have been people, and much of it has been intimate to their religions. Like other forms of art, music involves the extraction of order from chaos, and the capacity to do that is a reliable sign of life. People have long discerned that connection intuitively, making art essential to living. It is a connection, however, that deserves a more rational exploration. Sunday morning's service will give us several opportunities.

* * *

A PLEASANT SURPRISE greets me in the vestibule on Sunday morning: on the cover of the day's program is reproduced a small writing of mine, entitled "On Rainbows." I had written this in response to a call for essays on the subject of "forgiveness":

> My mother told me—as I suppose most mothers did then—that rainbows were God's promise not to repeat the deluge: a sign of forgiveness of human error. We have a rainbow of our own in the White Church on sunny mornings, cast in fragments about the sanctuary by the prism hung in a south window, the parting gift of Kirsten Lundblad. It is forgiving indeed, a small sign of grace in a place of sharing.
>
> A prism breaks the sunlight into its constituent colors by discrimination among wavelengths. Thus also are the people of Earth divided. Yet the breaking is illusion: light is one, and so is life. The parts are beautiful, and individuality expresses freedom; but the boundaries in the spectrum are indistinct. Nature offers us no sharp distinctions, but insists upon continuity and connection.
>
> Light forgives the prism and remains whole. The rainbow reminds us that life forgives our divisions and grants us each other.

If this is true, can we see beyond the metaphor and understand what we mean by life? Can we put it to a test, discuss it without darkening ambiguities? Where does it reside? Is it localized or shared? Does it have inherent boundaries, spatial or temporal? What other forms might it take? What is the distinguishing characteristic—if there is one—by which an inter-stellar traveler might recognize what we consider "life"? Would it be movement? Complexity? Self-replication? Are we content that our own view of life is comprehensive?

Many eloquent attempts have been made to capture in a few words this concept called "life." Some of these attempts remain cogent today, as reviewed by Barrow and Tipler in their comprehensive treatment of the Anthropic Cosmological Principle. [2] Most definitions include self-replication and continuity through natural selection among the necessary conditions of life. The subject, however, keeps escaping from definition as our world-view—and our introspection—evolve.

I am looking for something more fundamental, an understanding of "life" that would be recognized anywhere in our universe, even in other universes, a universal, intrinsic concept, entirely free of earth-bound biases.

Perhaps our attempts at definition are reaching for too succinct an answer to too many questions rolled into one. Suppose we were to set aside the puzzle of when complex molecules became "organic" in the earth's evolution of carbon-based DNA—which probably has a fuzzy answer—and focus on what we mean by *intelligent* life. Although obviously related within our human construct, that is really a different question, and might lead (if freed from anthropomorphic expectations) to some surprising answers. Then suppose we were to ask ourselves not for a definition (which is necessarily limiting), but for evidences, for characteristics, for how we might recognize intelligent life if it were to cross our path. In particular, what are the signs of intelligent life that would satisfy the scientists at the Harvard observatory?

Such evidence would surely include the capacity to organize, to develop or recognize coherent purpose out of random events, to affect the outcomes of external processes, to synthesize or assemble or build, to discern and reach for order. There may be many "life forms" with such a capacity in a very large universe, and not all need be carbon-based or even physical as we recognize physique.

Order is not enough; the universe is full of orderly crystals and large-scale structures that do not answer our questions. Coherence is not sufficient: the radio telescope at Harvard listens all day to invariant, continuous radiation from excited atoms at various temperatures, all quite coherent in their monotones, conveying no information at all. Even information content will not do: the more random and noisy a signal, the more information it carries,* but to no purpose.

We would need, rather, simultaneous evidence of (a) the capacity to recognize and bring order out of random, noisy, chaotic processes; (b) the freedom to choose such order; (c) the ability to communicate it in some form; and (d) the proclivity or intention to do so, an indication of purpose.

Only such evidence, we might agree, would be conclusive in the search for intelligent life, and it is a signal conveying that kind of evidence for which the antenna on Harvard's hill patiently listens. (Any of my four evidences of intelligent life is a necessary characteristic, but not sufficient; all four must be present to qualify. In contrast, being a human on planet Earth is sufficient, but not necessary.)

One should note that such evidences of intelligent life do not necessarily depend on all the requirements of human evolution; for example, on liquid water, or oxygen, or moderate temperatures. Nor need intelligent life be found only in creatures that see or hear or move about the way we do. Even the scientists who talk of finding life elsewhere speak in such terms, because our kind of intelligence

---

* Mathematically, the information content in any transmission is proportional to its uncertainty: a message known in advance carries no new information. This has been quantified in modern information theory (see the following chapter), which shows that the maximum "information" is carried in a transmission of perfect uncertainty—pure noise.

is the only familiar one. Such imaginary prerequisites, however, constrain not only our expectations but our search. We narrow our outlook severely by limiting it to worlds like ours and creatures like us.

The proponents of "Artificial Intelligence" (AI) have adopted a similar view, freeing the concept of intelligence from the particular necessity of a human brain. This has provoked wide and sometimes passionate objections, as it seems to propose an "unnatural" possibility of machines thinking and even feeling. In *The Emperor's New Mind*, Roger Penrose examines AI developments and limitations more dispassionately in his argument for a "non-algorithmic" element of consciousness. [3] Most of the philosophic heat is generated by attempts to draw a bright line between real and artificial, or between living and non-living, or between understanding and merely calculating, or between emotions and mechanics, etc. In fact, the universe presents us with *degrees* of living: the capacities to organize and communicate, and the freedom to do so with some purpose, vary widely among us and within us, from time to time and place to place, and are not perfectly achieved in any of us.

In the meantime we look for life here, for those evidences of intelligent life that sing to us some nearer songs. Our hypothesis is that intelligent life would be recognized by its capacity to discern order and effectuate it, including the freedom to choose; and by its ability and intention to communicate. From these fundamental abilities could evolve the capacities to learn and to build; to heal; to see beauty and make it; to love.* These are the songs of intelligent life for which we seek a rational base and a moral compass.

* * *

---

* The same fundamental abilities support what we call abstract thinking, creativity, memory, language, and emotions, and the crucial capacity to adapt to environmental changes. See Pfeifer and Scheier, *Understanding Intelligence*. [4]

THE REVEREND NATALIE WELLROCK[N] raises a newly-baptized infant above her head for all to see, and says "See what love has given us." That is my favorite sermon. But who is this little person? Is her identity already intact, or just beginning to form? (And who am I, then?)

Natalie talks about the church's hope to be an embodiment of the whole community, and thus to offer a wider identity to those who choose it. To put on this identity is to take on a shared self, an identity that does not get in the way of acceptance. Such an identity would be strengthened by sharing instead of by distinction. Natalie says to choose this identity, and it will make us free.

So perhaps I am someone, not in comparison to others but in identification with others. We are more accustomed to identity by distinction—strengthened by superiority, weakened by inferiority. Can we find ourselves rather in association? My worth, after all, should not be diminished by others' achievements or credentials, but enhanced by their sharing.

* * *

A HYBRID SCIENTIFIC/PHILOSOPHICAL literature has been inspired in the late 20th century by the implications of contemporary theoretical physics. Many of those immersed in seeking a consistent theory of how the universe really works have observed that they were also asking, in some sense, who we really are. Some have tried to put their discoveries in artistic and ethical context, or explored their parallels to mystical or religious thought.

Popular (although not necessarily representative) in this last category is Fritjof Capra, a researcher in high-energy physics at a number of U.S. and European universities. He speaks of our shared identity in his Preface to *The Tao of Physics:*

> I was sitting by the ocean one late summer afternoon, watching the waves rolling in and feeling the

rhythm of my breathing, when I suddenly became aware of my whole environment as being engaged in a gigantic cosmic dance. Being a physicist, I knew that the sand, rocks, water, and air around me were made of vibrating molecules and atoms, and that these consisted of particles which interacted with one another by creating and destroying other particles. I knew also that the earth's atmosphere was continually bombarded by showers of "cosmic rays," particles of high energy undergoing multiple collisions as they penetrated the air. All this was familiar to me from my research in high-energy physics, but until that moment I had only experienced it through graphs, diagrams, and mathematical theories. As I sat on that beach my former experiences came to life; I "saw" cascades of energy coming down from outer space, in which particles were created and destroyed in rhythmic pulses; I "saw" the atoms of the elements and those of my body participating in this cosmic dance of energy; I felt its rhythm and I "heard" its sound... [5]

Capra's book, and the later *Dancing Wu Li Masters: An Overview of the New Physics* by Gary Zukav [6], provide popular and accessible arguments for the consistency among liberal Western theology, Eastern mysticism, and the philosophical implications of modern physics. This doubtless speaks some truth, and their sharing of an integrative instinct is compelling, intuitive, and internally consistent. But is it more metaphor than science? Leon Lederman, a Nobel laureate and director of the Fermi National Accelerator Laboratory, dismisses such authors' more extravagant conclusions rather derisively in his entertaining and authoritative history of particle physics. [7] Yet his book also, and all those to which I refer you in these letters, is full of philosophical wondering (I did not say wandering.) To me they constitute in fact the most important development of philosophy itself in our age, and give us essential grounding in our search for a rational faith.

Thus I feel the tug of a mooring in the safe harbor of Natalie's Parish. I would offer you a rational translation, Mo, one that will welcome advancing knowledge, not fear it. Can we make our faith fully rational, and our search for truth entirely faithful? Should they not be the same thing? Can we make an unconditional commitment to truth that yet embraces the sense of beauty and sharing we find in moments of Bach and baptism?

* * *

I THINK LIFE has other songs to sing to us.

Notes

[1] Abby's story and her lesson on the meaning of life are given in my letter from Dorchester, Massachusetts.

[2] John D. Barrow and Frank J. Tipler, *The Anthropic Cosmological Principle* (Oxford University Press, 1988), 511–523.

[3] Roger Penrose and Martin Gardner, *The Emperor's New Mind: Concerning Computers, Minds, and the Laws of Physics* (Oxford University Press, 1989.) See pages 3–29 for a review of AI claims and tests.

[4] Rolf Pfeifer and Christian Scheier, *Understanding Intelligence* (Boston: MIT Press, 1999.)

[5] Fritjof Capra, *The Tao of Physics* (New York: Bantam Books, 1975.)

[6] Gary Zukav, *The Dancing Wu Li Masters* (Berkeley: Shambhala Publications, 1975.)

[7] Leon Lederman, *The God Particle* (New York: Bantam Doubleday Dell, 1993), 190–191.

Radio Communications Corporation[N]

L ABORATORIES ARE USUALLY good places to be. On the way from theory to practice, from mathematics to measurement, they are the engineer's womb, following conception and preceding delivery. They are sufficiently isolated, from the unforgiving environment where their products must eventually make their way, to grant us the time and freedom to make the best of our fragile ideas. Yet they are real enough to provide rehearsal in confronting, one by one, the difficulties encountered by ideas when thrust into practice.

At RCC our little lab is in the basement, a short walk from my office, down the stairs and past the coffee table, two rooms full of shelves and benches and racks of electronic equipment, all connected in a wilderness of wire. Here we share frustration, mistakes, shortages, impossible schedules, unrealistic expectations, equipment failures, and occasional success. Always there is the capacity to try, in a place where failure is tolerated: the freedom to be vulnerable.

* * *

THIS EVENING I am feeling vulnerable indeed. The technicians have gone home, the racks have gone dark, the day that started on Tilden Pond in Maine is ending in a Boston cellar. This morning my walk through tall pines to the lake shore was intercepted by the alien ring of a telephone. Could I return immediately from vacation? The coding demonstration, scheduled in a few days for our government contract officers, is not working and may have fatal design flaws. We need to decide how to correct or explain or....

So I left all my songs in the middle of their singing—your mom's sweet voice, your little siblings in the sand, the gentle embrace that the wind makes of pine and the waves of shore—and rushed back to the stress of clocks and traffic and competition and risks.

Now the bench feels unyielding under my touch, like the circuitry beside it, hard and foreign and out of reach. Facing failure, my mind turns away and replays past events: the excitement of building these benches, purchasing piece by piece the test equipment, seeing our first designs take gradual form in aluminum chassis, applying for patents, publishing papers in obscure technical journals: these things get funding, attract talent away from other pursuits, make careers. Our special-access, top secret security clearances are worn like the robes of a priesthood, and we preach the seductive religion of ever more complex weapons, countermeasures, signaling and cryptographic systems.

But it's not morning any more. The rough bench presses up on my forearms, the meters and switches stare down at me, pinching me between them for decisions. Beyond their problem, my stomach reminds me other things are not well. Why do we design things we know won't work in the field? How can we defend procurement practices that award contracts based on minimizing personal risk and pretending that the perennial overruns cannot be anticipated? Why should we be proud of being abstruse, especially those of us in the communication sciences? Is this a fraud, after all? Will my work reduce the world's pain, or add to it? What example am I setting for you, Mo?

We will solve tonight's problems tomorrow. But the older problem sticks to our corporate purpose. A better use of our training should be supportable, and I am resolved to try it. Matt's father is tired of this basement, too, and in our talks a new organization is taking shape, one of which you may be proud.

* * *

BEFORE WE LEAVE the lab I want to tell you something about communications, on which we spend all our time. I will not write so much about its technology as about its meaning (a subject so vast that my ineloquence may escape notice.)

Communication at some level is essential to all life; it is our means of conveying truth and sharing beauty; it is the infrastructure of love. Therefore we had better understand what it is. If we can learn to communicate more efficiently and more faithfully, perhaps we shall be able to comfort those in despair at Emerson Hospital. Perhaps also we may learn to hear signals of intelligent life on Harvard's hill, and understand why we sing Bach and what makes beauty. Perhaps we may find a way to reconcile freedom and diversity with sharing more of ourselves, extending our reach and our identities. And perhaps we may find a way to treat the suicidal strife that curses our species, find self-interest in other-interest, find what makes love rational.

There is, in fact, a science of communication, whose principles are valid wherever information is to be transferred. Although developed to enhance radio and telephonic links, it is applicable to more universal media—for example, artistic expression, organizing governments, organic communications among the cells of living organisms, discussion among individuals, and links within computers and large information-handling systems. The same principles would be valid for other life forms and other media about which we are so far ignorant.

* * *

FOR US IN the lab, the science of information theory begins with Claude Shannon. In July of 1948, he published the first installment of his seminal work [1], and most of what we do in communications, radar, and other signaling technologies builds on Shannon's foundation. Not intended as a philosophical discussion, the work is narrowly focused and mathematically rigorous (although remarkably clear.) In fact, Shannon says in his second paragraph:

> Frequently the messages have meaning; that is they refer to or are correlated according to some system with certain physical or conceptual entities. These semantic aspects of communication are irrelevant to the engineering problem.

In spite of Shannon's disclaimer, the theory seems to me of great philosophical importance. It establishes a rational basis for examining choice, uncertainty, and content in exchanging ideas among intelligent beings. And it establishes entropy as a measure of information.

Since 1948 the second law of thermodynamics and the concept of entropy have found their way into more popular media, for similar philosophical reasons. This was recognized by Robert Wright in Part II of his book *Three Scientists and Their Gods*, [2], although Wright hastens past Shannon to more accessible subjects. The substance behind this popular appeal is worth exploring, and Wright explores away, in an entertaining and clear translation of some pretty recondite concepts.

In the remarkable course of his book, Wright assays such modest topics as The Meaning of Life, What is Information (and Meaning, Complexity, Information, and Communication)?, and God. He does such a good job with entropy, kin selection, artificial intelligence, and other translations that I am spared the task, which I would not have carried off with his grace. He tugs and teases and peeks inside and runs around the edges of his three (actually more like a dozen) scientists and what they believe, converging (more or less) on a cogent logic of "self-interest [being] equated with a larger order"—an equation which humankind seems to be gradually understanding and trying to realize. And he gives me a running start on relating a rational study of communication to understanding what we really do mean by life.

* * *

So (DEEP BREATH) here is how I see that relationship. One learns from the practice of communication sciences seven requirements or principles; these are the things that make communication work:

- Uncertainty concerning the next message
- Context shared by transmitter and receiver
- Rate of message transmission
- Fidelity of the medium carrying the message
- Freedom of the transmitter to choose a message
- Acceptance by the receiver
- Veracity

(Be patient with me for a few pages. I want to use these principles of communication in exploring beauty, love, and identity, as well as life itself. No mean task.)

* * *

Uncertainty. We exchange information only when there is some surprise, some uncertainty, in what we are about to hear; i.e., when there is some choice in the message about to be sent. Without uncertainty at the receiver and choice at the transmitter, no new information can be exchanged. An endlessly redundant message may be comforting for a while, but it is surely boring. Choice and uncertainty support creativity. Information content is entirely missing from messages that are entirely known in advance, and at a maximum when least predictable. Information theory quantifies the uncertainty of each symbol transmitted, and names this quantity "entropy." [3]

* * *

Context. But perfect uncertainty does not work either. A context is essential: we can recognize a very large, but finite, number of symbols. In the language in which I write, they are a–z, 0–9, and other conventional symbols such as pi, the natural log base e, Planck's constant h, the speed of light c, etc.—symbols whose meanings we who use these languages have conspired to define in advance. If the message begins to contain unfamiliar symbols, the exchange of information suffers. In communication theory, entropy becomes infinite when the message is "white" noise, when we see wherever we look only random patterns that cannot be correlated with anything familiar. Such a message carries infinite information, but it cannot be decoded (understood.)

Besides agreeing on the symbols to be used, the transmitter and receiver must agree in advance (to some degree) on the time and place of the transmission, the medium to be used (acoustic, electro-magnetic, etc.), and the channel selected in that medium. They must also share some common experience, history, world-view, and other premises in order to share information efficiently. All these together constitute what I have called "context."

Douglas Hofstadter offers a somewhat different viewpoint on context, dividing it into two parts—the "frame message" and the "outer message":

> The frame message is the message "I am a message; decode me if you can!"; and it is implicitly conveyed by the gross structural aspects of any information-bearer....

> If the frame message is recognized as such, then attention is switched to level (2), the outer message. This is information, implicitly carried by symbol-patterns and structures in the message, which tells how to decode the inner message. [4]

Hofstadter's treatment deals with the general problem of information transfer when no *prior* context is established, when the receiver does not even know in advance that a message is intended, and we have to get her attention first. The whole communication, including its announcement and its code, comprises *new* information. This would be necessary in attempting communication between life forms who have no prior experience with each other. In our earthly experience this is rare, and the only new information is the "inner" message itself. (That is, we expect a message, know where to look for it, and share some kind of alphabet etc. with the sender.) Hofstadter's "frame" and "outer" messages have already been exchanged when we on earth begin most communication; they are what I call "remembered information," part of our natural shared context. [5]

Hofstadter sets this analysis in the middle of a broader treatment of "The Location of Meaning," in which he explores how information supports the conversion of DNA into physical organisms (and many less complex conversions), and how intelligence anywhere in the universe might recognize alien messages or life forms.

* * *

WELL, COMMUNICATION REQUIRES a *balance* between uncertainty and context. Consider poetry, for example. You would consider those passages to be beautiful, I think, that surprise you to some degree, perhaps with a word used in a new way, an evocative phrase, a passage that stirs a memory or demands a new look, a use or construction that arrests the reader and invites contemplation. Surprise is the measure of new information. Yet you will want the meter, the rhyme, the symmetry or sonority, or some structure or internal consistency in which you feel at home, and a shared experience to be stirred by metaphor. This is the "context." I propose that the same balance of surprise with familiarity is necessary in music and sunsets and relationships.

The simple Canon in D by Pachelbel, revived and popularized in the 1980s, gives me a musical example. It is charming on the first hearing, pleasant on the second, but eventually runs out of surprises and wears thin; we find less new at each hearing. Beethoven's symphonies last longer because of their greater complexity and surprise, but even they are less fresh after years of listening. In all of art, we find increasing content as we study the piece following our first encounter with it; then its content (surprise) gradually declines for us, over a length of time depending on its complexity and content; finally, with growing familiarity, the work communicates with reducing intensity.* The same is true of humor, which is often based on a startling communication, a sudden realization of connection. A joke tires very quickly because the surprise wears out: I've heard that one before.

Of course the truth revealed, the order perceived or discovered or structured, does not tire. The harmonies, rhythms, colors, patterns, metaphors, symmetries, sequences, are undiminished by discovery. The evocative power and the comforting effect of long-familiar works remain. In that sense, anything once beautiful is always so; although it cannot be revealed or discovered by the same person with equal impact repeatedly, it is available to others and to memory.

The balance between uncertainty and context is aided by the degree of redundancy or coding employed in our transmissions.

---

* I realize many will find this view of beauty invalid, even offensive. The argument is revisited, with the help of a most eloquent adversary, in the letter from Hard Scrabble Wash.

Languages are very redundant, such that a large number of symbols or even words can be randomly deleted before the text becomes unsure. (Shannon gives some examples in his paper.) Art is less redundant: a line of poetry is destroyed by an error rate that would barely obscure prose. (Some poetry and some music is intentionally redundant, but there the redundancy has a purpose of its own and adds to the communication.)

The arts communicate effectively because of this elegance; because they get our attention; because they supply their own mnemonics, they are easily remembered; and because they strike resonances in their audiences—resonances the artist knows to exist because of her shared context with the audience.

How many paragraphs of prose would I need to describe

> "...a leaf that lingered brown,
> Disturbed, I doubt not, by my thought,
> ...softly rattling down"? [6]

Frost knew his readers would be familiar with oak leaves in New England, hanging through the winter after other leaves had fallen, long since turned brown, falling at the least provocation. He knew we would remember the sound and the silence of a dry leaf, spinning and rocking through the air to the ground. "Softly rattling" is perfect: it is a novel use of the words, improbable until heard, and its surprise is enabled by the experience we share with the author, by the context. Frost has balanced surprise with context to connect with his reader. I could fill pages of descriptive, scientific prose without communicating as much. (This is not a threat.)

Shakespeare's line from Romeo and Juliet, quoted by Bronowski in *Science and Human Values* [7], provides another example:

> When Romeo finds Juliet in the tomb, and thinks
> her dead, he uses in his heartbreaking speech the
> words,

> Death that hath suckt the honey of thy breath.

> The critic can only haltingly take to pieces the single
> shock which this image carries.

Bronowski's point is metaphor; mine is the balance of context with entropy. The line sings to us for two reasons carefully balanced by Shakespeare: first, it is full of surprise, in the novel and multiple uses of words and images (see Bronowski's analysis); and second, it is set in the contexts of meter and metaphor to which the listener can be expected to respond.

The lines of poetry strike therefore a *resonance* in us, in the same way that a bell or crystal or tuning fork resonates when its frequencies are heard. We know in advance the antecedents of each metaphor and we are conditioned in advance to seek out the structure.

The radio telescopes' problem in seeking extraterrestrial signals is a contextual problem: they do not know in advance where to look. What channel should be tuned to? From which galaxies? When? Our distant, unknown friends do not publish a *TV Guide*. While these answers are hard to guess, some aspects of the context can be anticipated. The familiar constants of pi, the speed of light c, Planck's constant h, the mathematical constant e, and so forth, are universal, and any advanced civilization (where our laws of physics hold) would have to know them. Certain channels, corresponding to the natural frequencies of radiation associated with different elements (hydrogen, to be as simple as possible), would be obvious choices. So some parts of a context—a shared experience, a common ground—are intrinsic to intelligent life.

This gives us the possibility of connection. It may be that a translation will be necessary, or a series of them. I cannot write in Japanese, but I have a Japanese-American friend who shares the context of English with me and the context of Japanese with my audience, and she enables the communication. Perhaps we shall find the same need—the same opportunity—in reaching intelligent life that is several steps removed from our present experience.

* * *

Rate. The arts are "elegant" in the scientific sense: they convey substantial meaning in compact form. Elegance is a combination of uncertainty (content), context, and speed, which needs to be considered next.

The rate or speed of exchange governs the amount of information that can be conveyed within the time allotted for listening. In Shannon's information theory, the entropy of a message in bits per second is the product of its speed of transmission and the average entropy (between zero and one) of each successive bit. We need some efficiency, some economy in the transmission and reception. Without a reasonable rate of exchange, we wait too long for information, and if time has any meaning that diminishes communication. In any event, an economical transmission and an efficient reception will exchange more information than a partnership afflicted with great pauses, errors, or redundancy, and will be more faithful to their intent.

* * *

Fidelity. The fidelity or faithfulness of the communication channel is determined by factors like noise, attenuation, and dispersion of the message in transit. No channel is perfect, as any experience in electronic communication quickly teaches. The same is true in organic and personal communication. Our relationships, politics, business management, education, and all other components of the general commonwealth suffer daily from imperfections in their chosen media of communication.

Again communication requires a *balance*, between rate and fidelity. Too rapid a speech, too great a flood of impulses in the nervous system or computer, too many notes per measure, overwhelm the channel. It must be widened, or coding used, to correct errors. A perfectly faithful channel can be constructed only at the expense of all speed. To be certain of perfect transmission requires infinitely long error detection and correction. Neither perfect channels nor infinite speed of communication exists in our universe; we must accept an occasional error to sustain an acceptable rate of information transfer.

* * *

Freedom and acceptance. Often we speak of freedom but overlook its importance in communication. The exchange of information depends on the freedom that can be exercised by the transmitter in choosing among both symbols and premises (context.)

To match the transmitter's freedom of choice, the receiver needs an equal breadth of acceptance. Philosophically, this might be called trust or willingness or vulnerability; mathematically, it is a function of aperture, bandwidth, and quiet in the receiver.

Choice is the transmitter's entropy, and uncertainty is the receiver's, and they are measured the same. In a philosophy of communications, one might say that freedom cannot long exceed acceptance. The potential for freedom is born of the courage to be accepting, open, vulnerable in a communicative relationship. Again a balance is required, of freedom and acceptance. The transmitter's degree of freedom must be matched by the receiver's openness, by acceptance, else the information intended spills over, unheard and unused. [8]

* * *

Veracity. Finally communication requires veracity—a commitment to truth by the transmitter, and trust in the receiver. False information does not support life, convey beauty, sustain love, or facilitate the advance of civilization. All depend on truth. And all, therefore, are susceptible to error—not only to mistakes, not only to lies, but also to the pernicious effect of bias. Bias creeps in to unbalance the entropy of transmissions, to distort the context in which they are interpreted, to degrade fidelity, to limit freedom, to narrow acceptance, to undermine veracity. A hidden, persistent bias is more dangerous than large random mistakes. Some of us experience bias more frequently than others—perhaps because of appearance, gender, abilities, preferences, or other stereotypes—but all of us are vulnerable, both as transmitters and receivers.

* * *

So we have seven principles to observe if we are to communicate. They are the same requirements we have placed on our recognition of intelligent life. They are inherent in our sharing, connecting, cooperating; in our civilizing.

* * *

Well, Mo, I started this letter in a laboratory devoted to improving communication and then imposed on you a pedantic exercise. But you see, we can risk vulnerability here in the lab, and we enjoy freedom; we share sufficient context to support creativity and elegance, in our exchange of information and our search for better connection. At our best, we develop the "habit of truth" celebrated by Bronowski.

But at our worst? What might we learn in a different kind of laboratory, where such qualities have been abandoned or lost? Suppose we were to spend our days in a more brutal company: could communication and a habit of truth make a difference? Perhaps we could learn there more about what sort of creatures we really are, and how we might improve, and what separates love from isolation and hope from despair.

Notes

[1] C. E. Shannon, *The Bell System Technical Journal*, "A Mathematical Theory of Communication," , 27 (1948): 379–423 .(One should recognize that much pioneering work pre-dated Shannon's paper, by many scientists and engineers who deserve more credit than they usually receive.)

[2] Robert Wright, *Three Scientists and Their Gods: A Search for Meaning in an Age of Information* (New York: Times Books, 1988.)

[3] Shannon adopted the entropy because it was waiting for him as he sought an efficient way to measure how much choice one has in selecting a message, or how much uncertainty one faces in receiving it; that is, a measure of the information content in the message. He was able to show that only a peculiar mathematical form could accommodate all the criteria imposed by the logic of communicating a particular selection from a set of possibilities, when all we know in advance is their respective probabilities. That form turned out to be equivalent to one already established in thermodynamics to measure entropy, which quantifies the level of disorganization in physical systems. Thus Shannon named his measure "entropy" also, which is a most appropriate name for a gauge of information, choice, and uncertainty. It is important to note that the two entropies are not just metaphorically related, but mathematically equivalent; in fact, one can arrive at the thermodynamic entropy function by extension of Shannon's information theory.

In thermodynamics as in communications, entropy is a measure of uncertainty. The more possibilities are offered, the greater the uncertainty—and the choice.

Intuitively we expect important mathematical representations to be "elegant," that is, to express a profound realization economically. That is true of Shannon's entropy equation

$$H = -k \sum_{i=1}^{n}(p_i \log p_i)$$

In communications, this is the entropy of the set of all the probabilities "p" of symbols from 1 to "n," where n is the number of possible symbols that might be sent. Choosing a logarithmic base equal to n and dividing by k normalizes the result between 0 and 1. If a message source transmits "m" symbols per second, its entropy is mH. This would be measured in bits per second if only two symbols (0 or 1) are used. The maximum entropy per second is equal to the rate of transmission if all symbols are equally probable and each choice is independent.

"K" is simply a constant to accommodate the units of measurement to be used, and the minus sign is required by the fact that we are dealing in logarithms of numbers (probabilities) that are all between 0 and 1. The function peaks at 1.0 when all the probabilities are equal, drops off as any symbol becomes more probable than the others, and reaches 0 when any symbol is certain.

Shannon of course chooses 2 for his "n," which not only simplifies the math but corresponds to the choice available to the simplest electronic circuit, a switch. The switch chooses between "on" and "off" (1 vs. 0), and all of our digital communication and computer capabilities have sprung from that capability, endlessly replicated at higher and higher speeds. One can represent any pattern or compute any solution by adding and subtracting a sufficiently large number of 1s and 0s in the right sequences. (This seems to be, by the way, how our human brains work. Neurons fire or do not fire, based on the sum of inputs they receive, and when they do fire they send off that "1" to other neurons, and so forth. One might expect other intelligent life forms to employ a similar strategy. One should remember, however, Penrose's caution against assuming an algorithmic base for intelligence—see Note #3 in the previous letter from the Arlington Street Church.)

[4] Douglas R. Hofstadter, *Gödel, Escher, Bach: An Eternal Golden Braid* "The Location of Meaning," in *Gödel, Escher, Bach: An Eternal Golden Braid* (Random House, 1979), 158-176.

[5] See the letter from Tilden Pond for a discussion of "latent," "actual," and "remembered" information.

[6] Robert Frost "A Late Walk," *Collected Poems of Robert Frost* (New York: Halcyon House, 1939.)

[7] Jacob Bronowski, "The Creative Mind" and "The Habit of Truth," in *Science and Human Values*, rev. ed. (Harper & Row, 1965.)

[8] Technically, the transmitter's and receiver's entropies would be equal only in the special case where both are dealing with the same range of possibilities and their respective *a priori* probabilities. Then the transmitter's entropy goes to zero when she makes her choice (composes her message); and the receiver's when he receives and comprehends the message.

The Development Center[N]

BRAKES AND TIRES squealing! Car door slamming! Heavy footsteps taking our office stairs two at a time! Here is Mario[N], out of breath, two days early for our appointment. Can we help him? The Narcs are right behind.

Sure enough, a second car enters our small parking lot, equally in haste, and a conference is spontaneously convened.

The Bedford police station in only a block away, and I will accompany Mario and his new acquaintances for the booking, and tomorrow morning to the district court for arraignment, and then sadly continue our meetings in the prison from which he was so recently paroled.

* * *

TDC IS ONLY two years old and less than a year out of my basement. It has won sufficient contracts to fill the second floor of this old house-turned-office with a handful of employees and (fortunately for us amateurs) some professional consultants. Like most small businesses, it is growing in directions unanticipated at its formation. Matt's dad and I thought we would manufacture new electronic devices and do some training of technicians on the side, perhaps in prisons or the inner city. But the side has become the center, and we find ourselves deeply involved in trying out alternatives to the high-recidivism cycle of arrest-incarceration-prison industries-parole-idleness-re-arrest.

We are trying various types of training, job creation by forming new companies, pre-trial diversion of youthful offenders, counseling by ex-convicts, transitional employment to ease welfare recipients into productive lives, drug and alcohol treatment, large-scale prison visits and follow-up by community volunteers, specialized training with college credit for guards and probation officers, inmate councils with coaching in shared leadership, prison art projects with teachers and support from Boston's Museum of Fine Arts, and sponsorship of a host of other peculiar ideas brought to us by inspired and unconventional dreamers.

How much of this will work? Sometimes we will succeed. Often we will fail, occasionally with a mighty crash. We will set up an auto repair shop in the city and negotiate a contract to repair state vehicles, but one night a disaffected minority of our ex-con employees will steal all the tools. We will incorporate a building maintenance company and negotiate cleaning contracts with downtown offices, one of which our employees will accidentally burn down. I will stay up all night with two inner-city entrepreneurs writing out the charter of a new business to employ hundreds of their neighbors, then find their names in the paper the next morning, sought for the murder of a competitor. TDC will have few competitors, for good reason.

But the programs that start here really will employ and counsel and train and place thousands of men and women from the margins of society. Some of our ex-convict employees will prove to be excellent counselors of young people in trouble, tough enough to get a valid urine sample and stop a fight and talk about love. We will spin off a dozen companies, some of which will survive and do remarkable things. And in the process a number of extraordinary young men and women will work here and go on to larger ventures, matured and more expert than their employer.

All this is in the future, and now I must think how to help Mario and why our help has been inadequate so far.

* * *

MARIO IS A fellow entrepreneur, which may explain why we put up with each other. This fall he is in the copper business. He works two shifts, stripping it uninvited from Boston roofs at night and trading in the commodity markets by day. The commodity he receives in trade is a more complex chemical that he tells me will "cure anything, including living." Other than this addiction, Mario is intelligent, charming, and physically imposing. He has been a good student in our electronics technician class at the Middletown[N] prison, but so far has preferred his copper in sheet rather than wire form while on the street.

Heroin is a common problem among our students, and I have learned to detect its use by huskiness on the telephone, by missed appointments, and by the tears of young families. Mario has a young wife and a new baby, living in a part of the North End of Boston not on the tourist maps. When I visit they seem committed to each other and to a better life, but I am not very good at psychology or social work, and they are not really telling me much.

Our consultants have published books on the subject, and we understand the addiction well enough. What we do not understand goes deeper in our society and in ourselves. The choice of artificial stimulation or sedation over a hard reality is made by most of us at one point or another. As chemical and electrical intervention becomes easier with fewer contraindications, more will make that choice more often. Our argument will not always be sustainable by the threat of criminal prosecution or painful side effects. It needs to be made on the strength of choosing truth and connection over ignorance and withdrawal. That is a hard enough choice for the healthy and prosperous; to ask it of those in pain requires a new argument: that self-interest is other-interest. Can we make it so?

I would like to know more about who we really are. Somehow those who have lost their identity may help us find ours. Mario is on his way to a place where identity is in very short supply, and I will follow him.

Middletown Prison[N]

TO ME, THE most striking character of this structure is its efficiency. On this matter you may indict me as deficient in observation and sociology, shallow or insensitive or mechanical, and I shall enter a nolo contendere; however, that is what impressed me the first day: how few human designs there be as efficient as a nineteenth-century prison.

Superintendent McBride[N] showed me himself. From the central tower, one—I mean, one—can look down the East Cell Block, the West, the South, down the long rows of identical cages, stacked up four cages high, sighting down the long gray railings along the gray steel walkways that pass the identical locked doors, and see any door that could be open, any hand thrust out between bars, any forbidden item, any disturbance. Superintendent McBride informed me that a single man can throw the bolts on all cells at once from this spot, that in an emergency it takes only one guard to observe and control a whole prison. I was impressed.

Now I have spent hundreds of hours beneath the railings in the common areas, eaten in the cavernous mess room, worked in the industries, taught in the classrooms, visited in the visiting room, counseled in the offices, played on the exercise yard, even given a speech in the theatre. (There is no audience as great as a prison audience.) I am too familiar to bother with, and that is how I got in trouble.

This morning I drove through the "vehicle trap" to deliver my first load of tools and test equipment for our technician training program. Like the regular trap, through which all secured visitors and new inmates must pass, this is a double wall that opens one side at a time: the outer gate admits the visitor, then closes behind him before the inner gate opens. The intervening time is allowed for searches and other security provisions. My Chevrolet Suburban was full of tool boxes and cartons this morning, which the guards had been told I was allowed to bring in, and they viewed my cargo from only the trap tower above. So I drove on into the yard, backed up near the door leading to the classrooms, and recruited some inmates to help me carry in our equipment.

Now I am sitting in the Superintendent's office, recovering from a stern lecture on the inadmissibility of drills and soldering irons in a maximum security prison. The lecture dealt with such enduring principles as Trust and Responsibility, and was rather less remarkable for information than for redundancy. Correctional Officer Ryan[N], who discovered my offense, has made his triumphant entrance bearing some of the more offensive evidence, and has gone off to share the epic with his fellow officers. Fortunately my contrition and assurances have prevailed, and I shall be granted the privilege of recidivism—minus, of course, the tools of my shadowy trade.

* * *

So Mario has failed again. He leaves court in a familiar van, bound for the prison where I first met him, back along the road he and I drove together on a happier afternoon, back through the double gates and the trap and the body search, back into the "New Line." He has failed his young wife and child, whom he will not see except on Sunday afternoons in the visiting room, crowded with other failed families. The schools he attended so briefly failed him, the church failed after his first communion. Mario failed in the community of the North End and in the crowd of family and friends. Now he has failed at crime.

I look in his face to read the making of failure. Its chapters are thickly documented upstairs in his file, long stories and complaints and proceedings, chapters written by others in Mario's book. I have spent hours upstairs, in Mario's file and others. I know that the thicker the file the harder our job will be. I know the recommendations, the diagnoses, the sentences. But I want to see something else, the why of it. Where is Mario so different from Dr. Keezer, from Rev. Natalie, from me?

The men who cannot go home from here think they know the underlying distinction, and speak often of it: they are "numbers"; all their special things and even their names have been shorn away. Each one personalizes his cell, within the limits of regulations, but out in the corridors, down the steel walkways, in the endless queues, at the long feeding tables, on the graveled exercise yard, along the prison industries' production lines, even in our little classroom, they feel stripped of themselves. Beyond their clothes and few possessions, the stripping goes to their persons, and not only by the body searches. They know something is lost, and it is the most precious of things. And although they feel it was taken from them forcibly, I think their grip was weakened long ago.

It is identity. How strong is my own grip, I wonder? This is surely a place to learn. I try to listen carefully to those who might know. To the superintendent, it is genetic: look at their families in the visiting room, Dave; they just aren't high-type people like you. To the guards, identity is set by behavior, in a conscious choice: look, I grew up in the same tough neighborhood, had the same bad breaks, but I stuck it out and they didn't; it's who you choose for your friends and whether you try to get it all the easy way. To the chaplains and the evangelical program sponsors, identity is found in commitment to a higher faith. To the therapists it is more complex but still highly personal, involving adaptations to stress and building a sense of self-worth.

So it is personal genes or personal choice or a personal savior or personal self-love or self-understanding. But maybe the inmates are on to something: that it's not all personal, that somewhere they lost the ability to relate, to communicate, to connect—with others, with us. That disconnect has truncated their identities at tight borders, drawn in closely and defensively. In their way, they know this has a lot to do with who they are now.

Those of us who can leave the Middletown Prison tonight differ from those who cannot quantitatively, but not in all ways qualitatively. None of us has been faithful in every event. All of us harbor regrets, memories of omissions and commissions, a wish that we could escape from the old person—unworthy, guilty, unfaithful, disrespected.

Here I believe the news is good. If identity has to do with commitments and connections, we have a chance. The universe, I tell the men, supports for all of us an opportunity to recognize errors, become a new person, and forgive the old person within. We can escape from who we were by becoming who we would be, in relation to others.

* * *

I have brought women in here for the first time, and volunteers for discussion groups, and trainers and counselors and community delegates to a new Inmate Council. But today I have a softball team with me, made up of our staff and some ex-cons whom we have hired to counsel our younger clients. The game is played on the roughest field I have ever seen, its grass long since worn away if ever there was any.

Midway through the game a member of the inmate team collides with our second baseman, himself a recent graduate and known for his temper. I hold my breath as they sprawl together on the hard dirt; one is Black and one White, and they both have histories of responding violently to less jarring encounters. The inmate is up first, leans over my counselor and kisses his cheek. The silent yard erupts in laughter, the day is saved, the game continues.

I am unreliable in the field but occasionally effective at bat, and today I hit a generous pitch over the high wall, beyond the dirt field, out onto a street that none of the inmates has ever walked. I am exhilarated by my feat but as the men watch the flight of the ball I know they are thinking of something else. There is another variable that must be fit into the formula of life and identity: freedom.

The Pentagon, Washington, DC

THE PENTAGON HAS five of everything. Five sides, five floors, five rings of offices separated by five interstices, and a little five-sided gazebo at the center, serving food to everyone up to the five-star generals and admirals who defend our five freedoms.* It is the Trinity of our age, raised by two. A city in itself, large enough in scale and population to support all manner of shops and services, the Pentagon can be reached by air, water, rail, road, or tunnel (five means, of course.) And we peddlers reach it indeed, in large numbers.

The Pentagon's real commodity is neither weapons nor tactics, but money. Today I am one of the peddlers who shuffle in and out each day, bringing to the mostly-civilian buyers our fearsome wares—simple things in very large quantity or complex things in experimental numbers or sometimes just ideas, in ones and twos.

In my satchel is mostly chutzpah. To a building bristling with radio communication antennae, I have brought a new type of radio communication for sale. I have carried it down in my head to a briefing room that dwarfs, in size and sophistication, the laboratory in which it was born. In thirty minutes I figure out how to use the multi-media presentation equipment, and in three minutes I have lost my audience. Naively I start my briefing with an exponential

---

* So there are four freedoms, or maybe six. Poetic license.

expression depicting the waveform we favor, and before I reach my pitch the expressions in my audience have gone blank.

Years later the shared-spectrum address-coding techniques on which we work will be in wide civilian as well as military use, but the profits will be made by cleverer folks than we. In the meantime we will eke out enough funding to shuttle in and out of various secret chambers for a few years, and I can admire the Pentagon's inner workings and contemplate the guarding of freedom.

* * *

YOUR UNCLE JON is engaged in that guarding also, rather more urgently, on the other side of the world. He has flown fighters and bombers, but now he is in a more fragile craft, flying "forward observer" for combat troops in Vietnam. It is among the most hazardous of missions, and he has already been decorated for exceptional valor. One morning, while still dark, his base comes under heavy attack. With his comrades scrambling for cover and defensive positions, Jon races to his small craft and gets it into the air. He finds the source of the attack, calls in artillery fire, and returns to his field, safe for the moment.

Uncle Jon will be more fortunate than many and will come home. The nation's greeting, however, will be muted, not so much by the controversy of fighting that war as by the failure to win it. Much later I will wonder at the heroes' welcome given returning military personnel from the brief "Gulf War," which imposed (on our side) nothing like the horror of Vietnam. We are obsessed with Winning.

This is surely a city based on winning, and on power; I'm not so sure about freedom. In later visits I will have business in the Executive Office Building attached to the White House, in Congressional offices, and in one of the world's most expensive law firms. If freedom is bought by power, this is the world trade center. Something, however, seems missing.

* * *

ON MY WAY home, I stop at a quieter spot near the Lincoln Memorial. Here, where great and ringing words have been spoken about freedom, I spend some time with a newer sculpture, that of Albert Einstein. He is rendered in brass rather than the marble all about, gazing down at a celestial map, holding a thick sheaf of papers on which the sculptor has engraved the scientist's most famous equations—elegant quantifications of the photoelectric effect, the general theory of relativity, and the equivalence of energy and mass. This work of art and science captures, it seems to me, the combination of compassion and understanding and awe that distinguished the great scientist's career. This, it seems, has something to do with freedom.

Einstein knew that truth prevails in every event. His commitment to truth, wherever it leads, made him free, and we are all beneficiaries of his extraordinary freedom of thought. Yet all remain captive in some way, irrespective of strength and fortune.

Einstein also experienced the victory of a lie, cast over a whole nation and beyond. Ignorance—born of fear, prejudice, greed, deprivation, or pain of every kind—ignorance triumphs among nations, on our streets, in our homes. But it remains ignorance. One can make a lie believed, repeated, taught, rationalized, acted upon, defended: but one cannot make it true.

* * *

WHAT GIVES US the courage to deny a warm ignorance and embrace a cold truth? Back at home, I write a poem to your mom for our anniversary. The end goes like this:

> But freedom needs the stretch of love, to go
> beyond the narrow self toward who we are.
> The warmth of sharing dares uncertainty,
> Dissolves identity's close-bounded walls,
> > allows the reach for truth.

Other freedoms can be bought:
expression, strength, association,
movement, acquisition, worship –
all except the one we need:

The freedom from ourselves.
This bravest step, this flight,
      this reach, this daring;

This leap from clay to heaven is love's gift.

Who weaves our tightest bonds? What is the final freedom we need? How can truth be warmed enough to dare an unconditional embrace?

"…At last (time) steal(s) us from ourselves away" wrote Pope. Shall we defend ourselves from time? Or may we find somewhere the final freedom from our selves? Bonds we weave ourselves are not imposed by time.

Until we can see ourselves in others, it seems we will need walls, and they will wall us in as surely as they wall others out. In our individual relationships, as in international, it is hard to remain open to possibilities when we suspect another may not respect us, or has reason to distrust us, or perhaps the evidence to condemn us. Frequently we create our own expectations of attack, arm for it, and fulfill the prophesy. We all live with some sense of rejection, perhaps of guilt, certainly of vulnerability. If we could wear the whole mantle of truth, we might feel safe; but our knowledge of our own flaws exposes our flank. Courage starts with the acceptance within ourselves of imperfection. From this may follow a commitment to truth and connection, and a looking—in ourselves and in others—for the person behind the behavior.

* * *

WE LIVE ON a piece of a Colonial farm called Iron Oak, from which the young drummer Francis Barker marched on April 19, 1775, with his neighbors Abner Hosmer and Capt. Isaac Davis to the North Bridge in Concord. Francis returned to our fields but Hosmer and Davis did not. The Captain's home still stands around the corner from ours, flying a flag that was unknown when young Francis took up his drum. We are fond of saying that these young men bought our freedom with their courage. So did your Uncle Jon, and many of his friends who could not return with him. Their courage is demonstrated beyond question.

The purchase of our final freedom will require, however, a different kind of courage, to dare uncertainty and vulnerability in the reach for a wider being. Can we rise to this challenge? Perhaps our final freedom has been hidden, all these centuries, in the same place as the warmth of truth: in the unexplored potential of radical sharing.

* * *

MANY KINDS OF courage are needed for many kinds of freedom. We shall be limited in our freedom (and our perception) so long as we look for truth only in the well-lit places. When I return home I have a committee meeting in an institution that will demand more courage than I anticipate, a place of darkness, from which freedom of any kind seems impossibly remote.

The Newfern School for the Retarded[N]

D R. ADAMS[N] TELLS the committee there is a "back ward" in every institution—the building or wing or section where the most severe cases are housed, cases with no outside advocate, rarely seen by the public. He does not commend it to our view but offers in the spirit of openness to take anyone briefly through. Somehow I feel it will help my work, and accept his offer. Everyone else declines.

The Smith Building[N] is near the edge of campus, over a rise and far from the highway. Before I can see it I can hear it: the building seems to cry a continuous moan, varying little in intensity, with individual sounds rising only occasionally above the general cacophony. As we approach, it is possible to resolve separate voices, but no recognizable language.

The intensity of sound rises steadily with the narrowing distance to the front door. Now our conversation turns from easier topics to the shortness of staff and the triage process that commits human beings to this ward. Dr. Adams senses the growing unease that I did not expect in myself, and stops to give me a gracious withdrawal option. This will not be easy, he says, only a few staff ever come in here, and it will not make a difference in my committee work. I tell him it may help me elsewhere.

As we cover the remaining ground, Dr. Adams gives me my instructions: stay immediately behind him and do not touch anyone else.

Behind the heavy door, the noise level leaps alarmingly. Now its constituents are clearly individual, but incoherent. The air is heavy with organic waste, impossible to cover with the familiar institutional disinfectant. I feel immediately ill and vaguely frightened. Dr. Adams pauses in the administrative area to inform his staff of our presence and intent. We move quickly through the scenes of institutional life that I have come to expect in our prisons and mental hospitals, toward a locked ward, indeed at the back of the building. He checks to ensure I am following my instructions, then pushes through the door, locking it behind me.

* * *

THE WALLS I see are human. On all sides the forgotten cases, lost to all outside contact, slump and crouch. Much of their skin is broken by the scratching that continues all around, the only pleasure attainable besides masturbation. Arms and legs are irretrievably twisted and dysfunctional from years of disuse. Deformities I have never seen are common—missing limbs, unformed fingers, incomplete faces, growths of all sorts. Those who might destroy themselves or others—or who may have offended otherwise—are tethered to cots. All are naked, and living in their own excrement. There is no natural light, only fluorescent tubes high up, harshly reflecting on the uninterrupted tile, controlled by an automatic clock.

But I am entirely unprepared for what happens next. Upon our entrance the fierce incoherent shrieking suddenly resolves into a collective moan. The walls rush toward us, reaching, touching. There is no way to shrink from the frightening contact. Dr. Adams speaks sternly, pushing through the crowd. I stay in almost physical contact with him, keeping my hands away from those who beseech my embrace.

In a moment we reach the other end of the hall, negotiate the door and its locks, and emerge into another world.

I am shaken by the experience, surprised by how threatened I was. And I am ashamed by how much I could not see.

* * *

Is THERE A whole person inside those misshapen bodies? Does such a person have a claim to identity equal to mine? What do we mean by identity, and where does it begin and end? In the dark despair of prison cells and back wards, such questions are most starkly confronted. Here is a place to begin one's understanding of being a person.

Where does our identity reside? At the most elemental level, our bodies are mostly vast empty space, populated by mass/energy particles in constant motion, undistinguished from the particles anywhere else in the universe. If we examine the chemistry of our particle arrangements, we find mostly hydrogen, oxygen, carbon, and calcium, some phosphorous and other elements, in which none of us differs from inanimate objects. At the molecular level, the first evidence of what we call "life" is found in complex organic compounds, but here too we are identical to other organic accretions. Only as the self-replicating DNA is examined does a distinction among life forms become evident, and even then only upon scrutinizing the genetic codes that govern specialization.

Yet we know that genetics do not alone determine aptitudes and personality, and have less to say about behavior. I cannot understand who the back-ward residents are, who you are or who I am, by analyzing what makes us up, however sophisticated my physics, chemistry, and biology. Where, then, is the "I" in me? Of course it lies, in part, in the way my brain has been programmed since

conception, by environmental factors combined with internal processes working on the genetic inheritance. But this leaves an enormous range of choice: my freedom to control behavior, to enhance
chosen aptitudes, to modify the "personality" by which others will
know me. And it leaves unexplored my relationship to others and
the universe. If there is any meaning in who I am, it must have to do
with that freedom and that relationship. I would like to understand
them better, which is one reason why I'm here.

* * *

THE WORK OF our committee is to bring selected inmates from the
Middletown prison to work with the more difficult residents at the
Newfern School. The theory is that prisoners will respond positively
to the opportunity to help those even less fortunate, and in so doing
contribute to the rehabilitation of both their clients and themselves.
If the program works, it will relieve an overworked staff of some
unpleasant tasks, improve the school's image, help some residents,
and perhaps reduce the recidivism rate among Middletown parolees.

As representatives of the governor and various interest groups,
we strive more against bureaucratic than clinical pathology, but
gradually approvals are obtained and guidelines written, and the first
inmates are bussed the ten miles from Middletown to Newfern.

They are frightened and clumsy at first, but over the course
of the spring and summer they increase in proficiency and number.
The theory seems to work. Hardened prisoners who have spent most
of their lives in failure and rejection, unaccustomed to giving or
receiving, always suspicious and often violent, are learning to bathe,
dress, lead and sometimes carry others. The committee receives regular reports of progress, and most of the opposition retreats.

* * *

It is a sunny afternoon and I want to see some of the results of our program. We stroll down a familiar walk toward the Smith Building. Over the rise we come upon couples slowly moving across the walks and lawns: our prison inmates leading and wheeling Smith Building residents. The residents are dressed and clean and clearly excited. The inmates are solicitous, gentle; sustaining conversations that draw only grunts as response. The transformation is amazing: more obvious in the residents, but to me more impressive in the inmates.

I have seen these men in counseling, in court, in fights, in sports, in training and in jobs on parole, and I have not seen them reached the way these new assignments have reached them. They are as hard and brittle as people get, hard to any touch, hardened to therapy and hard to engage in constructive work, but here they can risk a softness. A soft breeze ascends the slope, and I realize my cheek is damp.

The Fitness Club^N

I N BOSTON, ONE does not get things done quickly through the formal channels of government, but by contacts with the unpublished infrastructure of campaign workers and influence traders that pervades the city's organizations after several terms in office. Even to get permission to play ball on any of the municipal fields, one must seek out the right member of the mayor's team, who may occupy an otherwise undistinguished position anywhere among the countless agencies of municipal service.

Although some of the mayor's functionaries are difficult to locate, Brutesto Sterone^N cannot be missed. To enter the Fitness Club, one goes down a flight of stairs from a homely square in the financial district, through the boxing rooms, Sterone is there, making the speed bag sing its triplets. He fills the whole door frame as I enter.

No, there are no more fields for practice or for games, but I and my friends might try out for some of the teams already admitted; have I played the game much? I square my shoulders and reach back for my best credential: for the first half of one season, I led the Belfast, Maine, Men's Softball League with a .500 average. Sterone is unimpressed. There is a team, however, called the Gray Grizzlies or something, for players over 40, that might have some openings. I cannot think how to get around the bulk of bureaucracy or its representative, so I mutter a thanks and cross the track, dodging Anna Robic^N on her 138th lap.

Mr. Thwok,[N] the tennis pro, is waiting for me upstairs, anxious to point out some of the deficiencies that Sterone and Robic have not already exposed. Thwok does not sweat, never needs to comb his hair, breathes pure oxygen, and arrives by some mystical prescience at the predestination of every ball I hit. He is obviously a Calvinist, and I resolve to consult him on my chapter on freedom, if I ever finish it.

Unhappily I do sweat, and breathe hard and fail to anticipate, and I am soon ready for a shower and the chilly walk back to my office. Robic is still on the track, probably up to lap 379 or so, now scowling at the presence of E. S. Cargot,[N] whom she must pass every 24 seconds. Cargot fills two lanes and moves at a...well, deliberately. To one side, Sterone is now working on the free weights. He asks if I'd like to "throw some around" with him, and I decline, fearing I might fail to dislodge them from the rack. In the center, Harvey DuPois[N] is laboring on a high-tech exercise bike, generously contributing to the universe's entropy. The bike's computer tells him everything about his time, speed, heart rate, degree of difficulty, calories burned, progress and prospects, but it will not tell him who he is.

Who is he, indeed, and who am I? All of us like to say we are here to build our health and character, but there is a large element of vanity in our motives, and something else. I would love to have Sterone's strength and Robic's stamina and Thwok's agility, and presumably Cargot and DuPois would even settle for mine. Would we then be satisfied? I think we are trying to become more attractive not only to others, but to ourselves, and the club will not do it for us, any more than the hospital will make us healthy, the church holy, the laboratory certain, the prison innocent, or the Pentagon free.

Those who frequent the Fitness Club come from many careers and interests, bringing with them many perspectives on our common quest. As we sweat and run and lift together, I listen to their stories and their hopes. All of us at the club have felt the quicksand on which we so often attempt to build the meaning and comfort of our lives. The structure is our identity, the way we recognize ourselves; and the unreliable foundation is accomplishment and appearance.

Sometimes accomplishment is perceived, say the members, in the testimony of others and can be objectively counted. Always it includes an accretion of expectations and perceptions and experiences, some defensible and some no more virtuous (or lasting) than beauty or charm or fame or wealth. Accomplishment is both positive and negative, past and future, personal and associative. We build our identities on failures as much as on successes, on expectations as much as actualities, on our influences as much as our accretions.

Over the years at the club it beguiles us, however accumulated, into a self-image that governs much of how we feel and how we behave, and that is ultimately very fragile. Beauty fades, limitations are recognized, time runs out on hopes. The structure of identity trembles or sags or distorts on its unsound base.

* * *

WE ARE ALL members of the club. Behind the question of "Who am I?" lies the pains of a trinity of identity crises:

- Who do others say I am (especially those whose opinions are important to me)?
- Who can I believe I am?
- Who am I in the sight of some absolute judgment or standard of truth?

THESE QUESTIONS DEAL with our experience of acceptance, internal and external, and with our understanding of what is ultimately demanded of us by life.

The question of Who am I? is usually answered by a recitation of training, profession, rank, family, residence, and influences (and subconsciously by an inventory of appearance, skills, and accomplishments.) This may be a happy answer or an unhappy one, depending on one's self-image at the moment and on what we think others may admire; but whether happy or unhappy, internal or external, it is neither a full answer nor a lasting one. The club members know this, for all their pruning and preening, and it haunts them. They would like a fuller and more permanent response, and one they could believe.

On a clear night after showering, when the lines of city towers converge in the stars, we tell each other that no one's deeds are mighty on a universal scale, no one's intellect towering, no one's accumulations significant:

> Then I looked on all the works that my hands had wrought, and on the labour that I had laboured to do: and, behold, all was vanity and vexation of spirit, and there was no profit under the sun....And how dieth the wise man? as the fool. [1]

What survives of identity, what still speaks of the worth and meaning of one's life when appearance and accomplishment fail? Around the track and on the courts and in the weight room one sees much of another quality: commitment. What if our commitment were to truth and to beneficial relation as a more secure basis of self-respect, not dependent on success, acceptance, or image? One cannot command success, perpetuate appearances, cling forever to unfulfilled hopes, or be certain of any outcome. But one can be faithful to one's commitments. Success will elude us frequently, regardless of effort. One cannot choose to be successful. One can choose, however, to be faithful.

We club members understand commitment, all right, if sweat is any measure. And if we think about it, we understand truth as the intrinsic reality in a non-capricious universe, that which prevails in every event, which passes every test; the integration of all information. We could explain that one can be committed to truth without ever capturing all of it, that one can distinguish it from error in particular situations without being able to define it, because to define is to limit. And I think most of us would agree that such a commitment therefore involves seeking more of truth as well as being truthful (as best we can discern truth) in each situation.

* * *

WHERE WE MIGHT disagree is in what we accept as a "given" truth, beyond questioning. My suggestion that nothing should be free of test, that a faith can be entirely rational, that we should be prepared to give up our most cherished assumptions and constructs if they are found to be in error, strikes the members as a harder discipline than all our bench pressing and racing and hitting. Yet that trust in truth seems to me essential to our commitment.

And what about a commitment to the benefit of others, to life? Can we agree on a rational test of what it is that reliably benefits life? We could start with our understanding of intelligent life as the capacity to recognize and bring about order (to organize, to build), combined with the freedom to choose our particular form of order and the ability to communicate about it. From these flows the capacity to learn; and thence to heal, to make commitments, and to see beauty.

Then one could benefit life by enhancing such capacities and freedom, and this would be a good test of faithfulness. Have we club members increased our collective order today, our freedom, our communication? Have we advanced the understanding and health and comfort and tools that may extend our reach? Have we taught or conceived or nurtured a child who is likely to live out the same commitment? Such questions invite a rational response to the broader challenge of finding meaning in the life we share, because they are based on a rational concept of life itself.

If we found at the club that these commitments gave us all a stronger, more lasting sense of identity, we would have found a greater security and strength, a meaning to cling to, a sense of worth. We could heal our feeling of not being fulfilled, of falling short, of dissatisfaction and ennui and that gnawing emptiness that comes upon us. We could trust in other-interest as self-interest.

We at the club are pretty good at commitment, actually; given a course we can trust, we'll sweat it out. I enjoy the company of Sterone and Robic and Thwok and Cargot and DuPois, because they care about something and stick to it, and I have the feeling that they care about me and would help if asked. But now I will ask them to reach deeper in themselves, for a harder element of identity. I will call it transparency.

* * *

We club members are here for reasons opposite to transparency: to become more visible, more noticed, more admired. Even after adopting a commitment to truth and beneficial connection, we still seek our identities—our self-images as well as our reputations with others—in appearance and achievement. We love the mirrors that cover the walls of the aerobics room, we love walking out of the weight rooms pumped up, we love an audience when we ace an opponent and pass another runner. It is the same with anything we fix or compose or earn: "Hey, my friend/spouse/child/parent/associate/constituency, hey, look at this!"

Transparency, however, requires getting out of the way, striving more to be looked through than up to. Transparency is achieved in contrast to the surrounding opacity. It is not a passive condition but an active achievement. Most of us prefer to draw attention to ourselves when we feel admirable, but this diverts attention from the message to the messenger.

I will not ask the club members to consider transparency just because the opposite is transient and distracting, however. I have a harder purpose, which involves reaching another level of lasting identity.

Suppose we consider what remains of King David or Beethoven or Einstein, or any of our heroes, few of whom presented a consistent example of ideal commitment or relationship in all respects. We see *through* them, however, things that were not visible before they joined us: these things are part of who they are, and now they become part of who we are. By the same process, our great leaders of religious, philosophical, and scientific thought were and are transparent to elements of truth that greatly enrich us that we did not see or comprehend before, that raise the place on which we stand.

Being transparent is not the same as being a "good example." An attempt to follow Beethoven's example is hopeless in his art (for most of us), and unprofitable in his behavior; yet we see better for the extraordinary window he opened. All our heroes are opaque in some respects while being marvelously transparent in others. Each of us will be found an unfaithful example in many respects, yet all of us have continuous opportunities for transparency.

I would like us at the club to learn to be transparent to such qualities and ideals that meet the test of benefiting life: that increase freedom, aid our building together, enhance communications and healing. What is revealed is learned; if real, it stays with us even when she who revealed it is gone, or found to be flawed in other beliefs or commitments.

Edwin Markham was a teenager when President Lincoln was assassinated, and might have observed the extraordinary grief that descended upon much of the recently-preserved Union following that tragic event. Years later he would find the voice to speak for all who held that memory, in a poem often quoted after the death of John F. Kennedy:

...he went down

> As when a lordly cedar, green with boughs,
> goes down with a great shout upon the hills,
> And leaves a lonesome place against the sky. [2]

MARKHAM KNEW HIS metaphor would have special poignancy in memory of Lincoln because of his physical stature. It is beautiful indeed, and consonant with the grief we feel when one who was much admired seems lost to us. It seems to me, however, just as errant on the point of transparency as striding in front of the club's mirror. We have needed heroes in the youth of our civilization, persons about whom we could construct elaborate myths and to whom great deeds and ideal characteristics would be ascribed by future generations. As we mature, we can recognize and accept the imperfections in those who have otherwise admirable qualities, and in ourselves. Such forgiveness seems to me essential to our identities as well as our health.

I do not wish to leave a lonesome place against the sky, because I wish not to be opaque to the sky before leaving it. The club members from whom I have most benefited were to me transparent—transparent to those lasting realities that command our trust and commitment; i.e., to which we strive to be faithful. When a member leaves the clubhouse, those realities remain; if she has been transparent to them, they are part of who she continues to be. And they become part of who I am, and who you are, if our membership is close and embracing. This is because our identities are stretched by each other.

This transparency is part helping us see, part teaching, and part sharing experience. My association with those club members to whom I have been especially close clarifies my vision, enriches my understanding, and extends my experience of life and truth. Their transparency extends beyond their borders as they appear in the mirrors. It affects not only what I see and learn but what I experience through association. This is their greater reach: they make a clearing in the fog that grows among us, through the walls that we build between us.

* * *

Is there something more to us, then? Do the identities of the club members, mine and yours, somehow include each other? Commitment and transparency imply relation and invite reciprocity. We look around at each other in the clubhouse, and wonder about the possible extension of ourselves, in each other. Why should there be a club rule, we wonder, that truncates our individual beings at some fuzzy extremities? It seems there should be more to our selves, in the real context of being in the universe, and it should deal with our connections, our giving and receiving, our communicating, our sharing of life.

We speak two basic words, says Martin Buber, I-thou and I-it, and the world is twofold for us accordingly. One can speak only the "I-thou" word with one's whole being:

> Whoever speaks one of the basic words enters the word and stands in it....Whoever says "thou" does not *have* something; he *has* nothing. But he stands in *relation*....As long as the firmament of the Thou is spread over me, the tempests of causality cower at my heels, and the whirl of doom congeals....All actual life is encounter. [underlining and quotation marks added] [3]

This is an investment. One may invest himself or herself in others, in effect transferring identity: the fullest gift.

I told Abby, before she died, that she would live in me, that I would carry in me the "real Abby"—her commitments, her relationships, the principles and ideas and qualities to which she was transparent for me. And this is true: the things I saw through Abby have become part of my life, and her identity lives in me.

* * *

WE AT THE Club know how to stretch. This is the stretch of identity: first to know myself, and be known, by my commitments; second to become transparent for others with respect to the realities that benefit life, to enable others to find new possibilities through me; and third, to stretch my self to include others, to live in *relation* to others.

Each expression of life is itself a matter of degree. The capacity and the freedom to recognize order, to build, and to communicate cannot be said to be fully present or entirely absent in any particular event, but its probability and intensity grow in proportion to the quantity and quality of relationships.

At the club we see the remarkable outward changes that can be wrought by work; it is not difficult to believe that whole identities change. The writer of this page is not the same person he was when very young, or will be when older; nor when ill, or confused, or asleep. He can be recognized by fingerprints or DNA, but only legally; mostly we rely on tracking the person through countless modulations, and pointing to the continuity of the track. I am closer to you, Mo, to you the reader, than to myself in a different state or at a different time—because we are communicating.

If identity depends at all upon interaction—even if we admit merely to a dynamic as well as a static component—then a larger identity is always possible. We do not see our whole selves in the large mirrors at the club. We are so committed to boundaries.

As individuals we are free to remain separate, or to join. The lessons of both history and physics are clear regarding Nature's preference for the latter choice. Those who cannot accept religious dogma as truth, nor subjugation to a will of God without rational grounds, can yet seek a wider self. They may seek it in relation, in commitment, and in transparency; and they may seek connections to a larger club and a wider membership than we have yet perceived.

The universe offers no absolute discontinuities. The nature of things depends on their relation to other things and to the observer. All things are connected, and all qualities are relative. The relation we seek at the club is not just the sum of contacts and influences, but the nurturing of such a reach that stretches those embraced by it.

At base, we are trying to find out who we *are*. Nothing in the universe—not the smallest elemental particle, nor the largest wall of galaxies—can be understood out of context, without understanding the influences around it and its interactions. One may say that nothing truly exists in isolation.

We do have what others need, for the greatest gift is to be truthful with them and to seek their benefit, in commitment, transparency, and relation. To look elsewhere for evidence of our worth is to miss the closest reality. This, I believe, is who we really are.

Does this sound strange, even mystical? Consider the electron, without which nothing we can see would exist. This very little character has been around since a small fraction of a second following the Big Bang, and he has become familiar to most of us since J. J. Thomson measured his charge and mass in 1898. The electron is known, however, only through his influence, only in relation. We cannot find him at all: he appears to be a true "point," that geometric thing that measures zero in all three of our dimensions.* Yet he has measurable mass, measurable charge, and tremendous effect. The impact of what I have called "relation" in considering identity

---

* As I write, the explanations of "string theory" appear more satisfying than the long-held "standard model" of fundamental physics. If true, concepts of point-sized particles will be replaced by vibrating loops of "strings" that have finite extent, perhaps in ten space dimensions. This will not change the observable behavior of electrons, or their metaphorical uses.

is just like that: an extension of an arbitrary position, an influence beyond limited boundaries, a reality outside a presumed location. This is more than a casual analog, it is the way the world works. (See the letter from The Board Room.)

* * *

My father's definition of a bore was someone who, when asked how he was, told you. So it is when we greet a new member of the club: we ask who she is, but we really don't expect an answer. Perhaps we should.

Perhaps we should say this: when we ask who you are, we would like to know what you believe in and why, what has meaning for you and how you perceive it, what you are seeking and where you look, what you trust and what experience you have with relation. We would like to hear of your commitment, what is important to you, how you see the world. We would like to learn how you try to be faithful to the best you can discern of life and truth, and how you try to discern more. We would look closely, not upon you but through you, to see what shines there.

And we will love you for that, because that is who you really are. No great achievement would make us love you the same way. No credentials, no accumulations, no appearance, no reputation will count the same for us. Our clearest vision of your person comes from seeing through you. We speak Buber's "Thou," and stand in relation.

Notes

[1] Ecclesiastes 2: 11–16 (King James Version.)

[2] Edwin Markham, "Lincoln, The Man of the People" (1901.)

[3] Martin Buber, *I and Thou*, trans. Walter Kaufmann (New York: Scribner, 1970), 54–62. (I have used the "thou" translation of the German "Du," because our English "you" does not convey the same sense of what Buber calls "relation.")

La Musée Marmottan

HERE IN THE City of Light we find the Musée Marmottan, adjacent to the Jardin du Ranelagh and a block from the eastern edge of the Bois de Boulogne, and here we contemplate, nearly to surfeit, the work of Claude Monet, who made the individual transparent to the light. Monet strove to paint the light, the "air itself" he said, and in so doing made it uncertain where "identity" begins and ends. He saw that the truth in identity did not lie within boundaries. Thereby we find the identity of the lily and the cathedral and the person enhanced, not diminished, in his images.

As Monet and his impressionist contemporaries worked, physicists were beginning to recognize the interconnection of all the universe, the irreducible uncertainty in measurement, the relativity of space-time, and the role of the observer in defining reality. Like Monet's representations, these are not limits of vision but consequences of being here.

We are always looking for distinctions, for precise boundaries, for definitions, beginnings and endings. We have interpreted our experiences to reinforce this discrete view of the world. But the real universe confronts us instead with continuities. Our boundaries cannot be pinned down.

Writing in *Natural History,* Stephen Jay Gould raises some basic questions about individuality:

> Nature is not an intrinsic harmony of clearly defined
> units. Nature is built by multiple levels, interacting

fuzzily at their borders. We cannot even formulate an unambiguous definition of individual at the single level of organic bodies—as Armillaria mats and aphid clones demonstrate. Moreover, in Darwinian terms, legitimate individuals exist and operate at several levels of a genealogical hierarchy—genes and species, as well as organisms. [1]

Where then shall we find the "legitimate individual"? An all-absorbing universe, without individuality or diversity, would be colorless, uniform, devoid of character and distinction. Can we find individuality enhanced by joining, by sharing? Is freedom compatible with a sustaining context?

* * *

FROM THE MOMENT of the universe's initial singularity, its average temperature has declined, heavy elements and complex molecules gradually evolved, and an inexorable cosmic process played out in accordance with no apparent command but probability. No choice is evident in this process, whose destination is an equilibrium of disorder. Yet in at least one of the intermediate aggregations, a freedom of choice has evolved. A power of decision appears; an ability to discern order from disorder, and at least locally, an authority to choose one over the other. This we recognize as an indicator of intelligence, of conscious life.

Life presents immediately a paradox: our freedom to choose is an individual matter, whereas our ability to make the choice work depends on the cooperative effort of a whole people—perhaps, in the end, on our embracing all life as one. Choices become progressively grander as the species and its civilization evolve, eventually embracing the possibilities sought from Harvard's hill.

So we arrive at a choice between union and fragmentation, between coherence and dispersion. We are free to seek or to rest, to reach out or to draw within. As a people, we are capable of great things, or nothing; of building up or of tearing down. This freedom to seek and question, and the openness to accept challenge, are essential values of a civilized society. They are also tenets of scientific inquiry.

The artistic quest for beauty and the scientific search for understanding, Bronowski says, require both a "habit of truth" and an exploration of hidden similarities. The advance of science, art, and ethical standards find their common base in that habit, that exploration, that commitment; as they advance, we can increasingly see a central theme which Coleridge defined in the arts as "unity in variety." [2] So the appreciation of great art, as its creation, is a discovery of order, connection, and unity in an infinite set of possibilities—the derivation of meaning from random events specially arrayed. That discovery, so evident here at the Marmottan, derives from our freedom to seek, our capacity to arrange—and our openness to new events.

Reconciling individuality to society suggests a revolutionary opportunity: that freedom may be increased by connecting. This has not been our way, at least in Western thought: fighting for liberty, distrusting external influence, we resist encroachment upon our separate selves—and we build our own walls. To breach such walls has always threatened us, because we are more afraid of intrusion than hopeful of escape. Are we ready at last to seek an opening that may lead us to a freedom not previously known? A wider self has the capacity to be freer than a narrow. Self-interest is other-interest.

Other-interest is self-interest. If we are free to join we are free of boundaries. Philosophers have long recognized the importance of freeing the mind from its physical fetters; but the individual mind itself needs freedom from its own limits. Bronowski quotes Alexander Pope:

> Years fol'wing Years, steal something ev'ry day,
> At last they steal us from our selves away...

Bronowski is here exploring "The Creative Mind," showing the thought processes in artistic expression to be the same as in scientific discovery. [3] Pope was lamenting how temporal even the most creative and hard-working mind must be. Both recognized intelligence as the essence of being, but both wrote in the tradition of an encapsulated mind.

In our more recent experience, we see the immense expansion of creativity through cooperative thought. Our advances proceed from many minds working together, sharing, learning from each other, building a common context that will support ever-longer reaches of discovery, communicating. The comprehension of unities among discrete components makes art and underlies scientific thought, says Bronowski, and I believe it to be fundamental to conscious life itself. In fact, I would propose that this capacity of pattern recognition, of running correlations, of detecting hidden connections, of integrating incoherent floods of information and stored memories until insights and discoveries rise out of the noise—that this capacity is the "nonalgorithmic" capability that Penrose finds lacking in "artificial intelligence" claims and that Hofstadter sees in the "tangled recursions" of intelligence.* This seems to me important not just as a hypothesis on how the human brain may work, but as support for the cooperative advances possible as we work and grow together. Such cooperation gives us a wider view, a larger memory, an enhanced ability to correlate and recognize, to discover and advance. It supports our "becoming together," which I have proposed as an element of the meaning of life. (See the letter from Dorchester, MA.)

This is not to devalue those individual discoveries that so often accelerate scientific progress, nor to dismiss the importance of

---

* See notes 6 and 7

leaping over conventional and popular convictions in one's search. There is no virtue in joining an untrue cause or participating in illusory or malignant comforts, however widely shared. Indeed a persistent theme in these letters is the unexplored potential of finding reality beyond commonly-accepted limits. The histories of art and science, and of civilization itself, are punctuated (as Gould says of evolution) by challenges to the prevailing order. The testing and gradual acceptance of such challenges leads to a higher order, only to be challenged again as our understanding advances. Monet himself achieved such an "individual" breakthrough in art, as has Gould in science. They must have felt very much alone at times. Yet their work rose from a base built by others before them, and has gone into the base on which we work.

When we say each other we imply a free association, not a surrendering of identity. "Each Other" is a concept of inclusion but not of absorption; not a loss of self but a larger self, in relation. Our becoming together, punctuated by contributions of each for the other, is the opposite of an equilibrium in which distinctions dissolve. It is the most improbable thing in the universe, the step-by-step building of new order. It is the exercising of our capacity to learn from each other, to heal each other, to love each other.

* * *

ARE WE "BECOMING," in fact, as a people, as our strength and understanding grow? How might we respond, for example, should alien beings invade the earth? The very words betray our orientation: what is alien? what is an invasion? Could we open ourselves to new possibilities? Could it be that we have no evidence of other life because we are not yet mature enough to perceive and accept it? Lewis Thomas, reflecting on the quarantine that astronauts undergo upon returning to earth, observes that many of our illnesses result not from an attack by our guests but from the inhospitality of their host; so it may be of the illness of our society:

> If there should be life on the moon, we must begin by fearing it. We must guard against it, lest we catch something.
>
> It might be a microbe, a strand of lost nucleic acid, a molecule of enzyme, or a nameless hairless little being with sharp gray eyes. Whatever, once we have imagined it, foreign and therefore hostile, it is not to be petted. It must be locked up. I imagine the debate would turn on how best to kill it.
>
> ...It says something about our century, our attitude toward life, our obsession with disease and death, our human chauvinism.
>
> There are pieces of evidence that we have had it the wrong way round. Most of the associations between the living things we know about are essentially cooperative ones, symbiotic in one degree or another; when they have the look of adversaries, it is usually a standoff relation, with one party issuing signals, warnings, flagging the other off. It takes long intimacy, long and familiar interliving, before one kind of creature can cause illness in another... We do not have solitary beings. Every creature is, in some sense, connected to and dependent on the rest. [4]

Thomas's insight reminds us that the process of connecting is necessarily reciprocal. I must be understood myself, by those with whom I would connect. To be understood, I must be truthful. I must show my real self. Thus I need self-understanding to be in good connection with others: first to learn what it is I seek, and second to show what it is they seek. The shared enterprise—the common under-taking, which can do so much more than individual efforts—requires mutual understanding to work. Truth is its fuel and lubricant.

* * *

WE CONTINUOUSLY RECEIVE signals from a very short range. At our most primitive, they are all we comprehend and we respond only to them. Their volume is increased by pain of all sorts. As we advance, signals from a greater distance may be heard. Those who raised the ground on which we stand bought us the freedom to listen.

We are better receivers together than apart. Like a phased-array antenna, a whole people can resolve a smaller signal at a greater distance, and present a wider aperture to its message. We listen with different ears, from different places, perceptive at different moments, tuned to different resonances. When we combine our hearings, a finer grain of truth may be discerned and a farther horizon scanned. That is the genius of diversity: a heterogeneous community is a healthy community, combining a broader range of perception and experience, thus enhancing access to more of reality.

Where is the reality here? Within the isolated cranium? I study my fingertips as I type: shall I believe that my identity ends at their extremities? Or are we ready for a courageous leap? If one is committed to truth, a leap of faith is a leap of rationality: no more, although that is great enough to make all the difference; and no less, although it is close enough to trust.

* * *

WE KNOW THAT the smallest things discernible behave sometimes as particles, and sometimes as "waves" that vary in intensity over time and distance, depending largely on how we choose to observe them. It is easy to believe this of the ghostly photon, that massless "particle" that carries the "wave" of light and other electromagnetic radiation. It is harder to swallow a dual identity for things with mass, like the familiar electron, but the experiments seem unambiguous. Even beyond the point-sized electron,* wave-like behavior can be demonstrated for atoms and things we can see.

These waves have a frequency that varies with the energy of their associated particles: the more energetically they jiggle around, the higher their respective frequencies. This means that their wavelengths are inversely proportional to their energy, so that as particles are cooled down the wavelengths associated with them spread out. Albert Einstein even predicted (with S. N. Bose, for whom the Boson was named, in 1925) that near absolute zero temperature the wavelengths of adjacent atoms would overlap, forming a new kind of matter in which the atoms lose their identity and cluster together in a "Bose-Einstein condensate." Recently this effect has been demonstrated in the laboratory, and may have important practical applications in the future.

In fact all matter—including us—has its waves, its variations of intensity, its uncertainties of exactly where it is and where it's going, its constructive and destructive interferences, its propagation of influence beyond its presumed spot. Nature is like that. Once we get our minds around the fantastic consequences of this strange duality in the stuff of the universe, it is not so hard to consider the extensive possibilities of our own identities.

* * *

IT IS SAID that Einstein conceived his great leaps of discovery in "thought experiments" before the mathematics were formulated. Perhaps his most famous thought experiment was supposing that he could ride a photon and experience its relationships. Suppose, then, that we imagine riding an entity of consciousness from another

---

* See note, p. 73.

world, carrying an "intelligence sensor," seeking companionship around the universe. Where shall we look, and what might we find?

Consciousness does not necessarily require an organic brain: that is almost a pre-Copernican conceit. It could presumably be borne in a cloud of energetic particles, as imagined by Fred Hoyle. [5] Or it might be found in other concentrations of storage, processing, and communication provisions making use of input sensors and output transmitters (which some would say characterizes the human brain, for example.) The point is that intelligent life is a matter of recognition, organization, freedom, and communication, and need not be limited to familiar forms. But would a very complex computer qualify, or do our brains have a more special capability, one that cannot be synthesized from digital circuits, however sophisticated? What, if anything, sets our brains apart from the machines that "artificial intelligence" (AI) researchers hope to develop?

Hofstadter and Penrose have both explored, more eloquently than I can manage, the "non-algorithmic" nature of consciousness as we humans experience it. To Hofstadter, the "essential abilities for intelligence" that our imaginary traveler would seek are:

> To respond to situations very flexibly;
> To take advantage of fortuitous circumstances;
> To make sense out of ambiguous or contradictory messages;
> To recognize the relative importance of different elements of a situation;
> To find similarities between situations despite differences which may separate them;
> To draw distinctions between situations despite similarities which may link them;
> To synthesize new concepts by taking old concepts and putting them together in new ways;
> To come up with ideas which are novel.

> …Without doubt, Strange Loops involving rules that
> change themselves, directly or indirectly, are at the
> core of intelligence. [6]

Roger Penrose suggests that one's consciousness can make contact with, or have a direct route to, intrinsic "Platonic" truth (at least the more profound mathematical ideas), and that this perception cannot be explained (or reproduced in AI) by the evolution of computer-like brains based solely on complex algorithms. He speculates further that the brain may organize itself and grasp profound truths by non-local resolution of overlapping (superposed) quantum states, as would a quantum computer, making the higher-level "conscious" functions distinctly non-algorithmic. [7]

These imaginative attempts to explain "conscious" functioning expand on my reference to the organizing/recognition capacity of intelligence and its requirement of freedom. They seem consistent also with the capacity to perceive hidden connections, to discover relationships, to use metaphor, and to strike "resonances" in communication (whether artistic or scientific), as explored by Bronowski.

So we are free to join, and increasingly free of boundaries. When we seek intelligent life we look for evidence of freedom, along with the capacity to choose order and communicate. The evidence must reveal both a recognition of order and a modulation of disorder (new information) upon it. Life announces itself by daring the embrace of uncertainty.

We are joined, you and I, in a common enterprise. We share the particles and energy of the universe and constantly exchange them. The capacities to learn, to heal, to love, and to build are collective capacities. Whatever our differences, we share common needs and potential. Whatever the barriers to our becoming, they will be overcome only together.

The "creative mind" explored by Bronowski works, in art as in science, to find relationships. Truth has been a cold comfort through the dark youth of civilization, feared as much for its uncompromising prevalence as for its elusiveness. Those who fear turn to magic, religion, and the arts for comfort; but truth prevails in every event. Is it not time to trust in truth and not fear it? We may do so, because science seeks the same connections as do art and civilization, and we can depend on those that are found. Our refuge has been to hide and to appeal to the supernatural through our dark youth, but the refuge of a mature people is truth. Science and art discern connections, seek unity, find order. Truth is warmed by the creative mind, and by sharing.

Bronowski's creative mind works in analogy and metaphor, whether in art or in science. It is the capacity to recognize such associations that characterizes consciousness—and the form of intelligence we hope to find beyond our own. This is the same capacity to organize, to discern (or create) order out of chaos. While recognizing the insights such associations inspire, rational thought must go beyond them to intrinsic reality. Indeed Bronowski points out the result of insight must be subjected to test for corroboration. He says these capacities and associations are functions of the creative mind; they are surely the signs of life.

* * *

WHAT, THEN, DO all those whose strivings we have witnessed seek? What is the common struggle engaged in by the artist, the hospital patient, the scientist, the worshiper, the prisoner, the retarded, the alienated and loving, the dying and living? We had better understand this, for we are they. Each of us is at times ill, imprisoned, retarded, guilty, grieving, frightened, unconscious, in pain—just as we are at times healthy, free, insightful, innocent, happy, secure, alert, comfortable. We are never far from the heights of power or the depths of despair. So what might we bring to those who strive?

Can we bring to them—to us—comfort in their dark and identity in their light? If we can comfort pain, we quiet the near noise that prevents hearing of farther signals; we free the listener to seek her identity. But the tight embrace of comfort can also stifle the growth of identity, while the seeking of meaning often reduces comfort. Can we resolve this conflict in relationship, in sharing?

The members of the Fitness Club taught us that the qualitative identity of each being is recognized by her expression of commitment, transparency, and relation. Which part is most obscured, and what obscures it? My candidate would be relation, obscured by the fog among us. It is that fog through which our transparency aspires, and that aspiration to which our commitment is made.

Besides comfort and meaning, people need a sense of the holy. If we take away one object, they will fasten on another. In rationalizing our view of the universe, we must not diminish the sacred, but enhance its recognition. The holy remains when the mystery erodes, like a limestone cliff, full of ancient life, shining forth upon the erosion of its sandy cloak. We have been inhibited by fear from shedding our cherished mysteries; but they only cloud the real substance of the holy within.

Notes

[1] Stephen Jay Gould, "A Humongous Fungus Among Us," *Natural History Magazine*, July 1992.

[2] Jacob Bronowski, *Science and Human Values*, rev. ed. (Harper & Row, 1965), 16.

[3] Ibid., 17.

[4] Lewis Thomas, "Thoughts for a Countdown," in *The Lives of a Cell* (Bantam Books, 1975), 6.

[5] Fred Hoyle, *The Black Cloud* (New York: Signet Books, 1959.)

[6] Douglas R. Hofstadter, *Gödel, Escher, Bach: An Eternal Golden Braid* (New York: Vintage Books/Random House, 1989), 26–27.

[7] Roger Penrose, *The Emperor's New Mind* (New York: Penguin Books, 1991), 428–448.

Shaftsbury, Vermont

I HAVE LOOKED THE pale sun down, low along the pines, reluctant, procrastinating, hiding in the winter ice-clouds, sliding at last behind the western hills. In his place the snow-squall growls from a great distance beyond the horizon. Her long rolling voice sweeps through the conifers, felt more than heard in those low tones that carry far. As she comes into sight, scrambling over the New York mountains down into our narrow valley, her voice rises in pitch and volume. The snow is black against a shrinking sky. To the north it races across the brief fields and climbs again, alternately hiding and revealing the Green Mountains' rise.

I cannot follow such a flight, and call her nearer. She turns south toward me, roaring down the valley, drawing darkness behind. The trees around me whisper, gasp, groan in chorus, swaying, anticipating in the great breathing phrases of their highest branches. My small space below is hushed in suspense. As the snow approaches it turns white and disaggregates into crystals—driving clouds and small cyclones of crystals, resurrecting their ancestors from the frozen ground. The storm's voice finds nearer resonances—baritone then contralto then soprano, as she invades closer things. Finally long grasses begin to bend across the nearest field, the bending rushes toward me, and the ice-wind seizes my face in the rough embrace of winter.

I have called that embrace to me, out from behind the sunset, to experience a sharper present. Alone in the Vermont winter one listens for a farther voice, personifies the storm. The icy grip alerts all the senses, straining to discern what one cannot find under shelter. There is a presence here, one feels, and my shivering is more than the cold. I remember Martin Buber's words: "Here the relation is wrapped in a cloud but reveals itself, it lacks but creates language. We hear no You and yet feel addressed..." [1]

It is enough: I cannot sustain the storm's embrace so long and so gracefully as the Canadian hemlocks: I leave them swaying in perfect dance with her, and retreat to the warmth of a hearth and human touch.

* * *

HERE AT THE winter solstice in Vermont we celebrate our family, almost all together, the snow falling outside warm windows beyond the lighted tree. A great pile of gifts is brightly wrapped beside the hearth. The children's laughter accompanies the music of a wood fire and a baroque oratorio. It is too pretty, and I look beyond the Green Mountains and hear crying.

My shivering has followed me inside. Until we can hear in the universe an echo of our love, until we can stretch to more universal reaches, our hearth must grow cold. Outside, beneath the snow, lie miles of stone walls, built along with our comfort by those on whose shoulders we stand. Beneath the snow they lie there too, and I have read their graves. Their pain and their progress survive them, and dwell unevenly among us.

As the storm swirls up its crystals, it raises also from the ground, from the builders of walls and fields, from ancient strivings, a song I cannot quite make out. It is a winter song, masked for the moment by our indoor brightness, shut out for a while by our walls, but I can hear it out there and I know it will still be there when our music stops. A fragment of it sounds like this to me:

> Who grants you grace? What larger part
> will shade you from your Star,

> that you discern a joinèd heart
> and who you really are?

The poem suggests that love and identity are incomplete as long as there is crying anywhere. It begins in the place where Robert Frost said all poems start, in pain. But you have asked me to go beyond beginnings. Can we say what we really mean by love, and who we really are?

A little way to the south, Frost's Shaftsbury home still stands, beside the fields and woods and hills that gave context to his work. Still the Frostian ambiance is palpable in all seasons here, providing still the shared context that makes poetry work. I do not need Frost's poems in Shaftsbury, because they still blow in the Vermont wind, walking the sodden pasture lane through gaps in stone walls not mended, across the snow by the edge of woods, bending the birches, whispering to the ground. Instead I bring Shannon's information theory and Buber's concept of Relation, and find them singing the same song. This is what they say to me:

* * *

A SHARED CONTEXT gives us not just poetry, but scientific progress and civilization and the experience of love—allowing vulnerability and supporting an efficient kind of communication that we call intimate. This experience ultimately is of unconditional acceptance, confirming and reinforcing our mutual identities. I propose that love, like faith, can be considered a combination of trust and commitment—a commitment to the benefit of another, and a trust in the other's equivalent commitment. And I think it includes something else: a sharing so close that one feels part of another, and experiences the other as part of one's self. I will call this "extension"—the enhancement of identity through the experience/perception of inclusion in others.

A rational exploration of something as important as "love" demands clarity in what we mean by each of its postulated constituents. Words are only symbols, of course, and suffer from both incompleteness and varying connotations. If one subscribes to the Platonic view that real concepts exist irrespective of our perceptions (or wording), then our use of words is an attempt to convey such real concepts as we see them.

Trust in this context, then, seems to me an expectation of truth as well as benefit from others. And the measure of "benefit," if it is to be real and lasting, is the degree to which the capacities of life are enhanced—those capacities being freedom and the ability to find or create order and to communicate, to learn and commit. One trusts the concept, object, process, or person from whom one expects such enhancement, and in whom one discerns a commitment to truth, unconditionally.

Reciprocally, commitment to others is a dedication to their benefit, to their realization of these life-capacities, and a dedication to be entirely truthful with them, that they may put their trust in us. Such a concept of mutual trust and commitment, I hope we can agree, would be essential to the advance of any community of intelligent life, here or distant. A community from which such trust and commitment is entirely absent would more likely destroy itself than build a lasting civilization.

I propose further that such a trust, underlying love and faith, supports the courage to be entirely open, to share, to risk mistakes, to build together, to listen and learn, to communicate. This makes civilization work, and science, and art. Moreover, by placing our confidence in truth and in life-enhancement, we are recognizing by definition the object of a rational faith. Trust invokes its own object, because it is an expectation of truth. To trust is to hold this confidence in the face of any risk, and to consider and behave accordingly.

Thus we experience love—through unimpeded communication, acceptance, courage to be open, broadening identity through sharing of ourselves, through extension. Thus also we give love, by seeking such a shared context for communication, by committing to connection and trusting it, by accepting others without condition, as part of ourselves. I consider this a natural order and the direction of evolution.

Together we build a context of trust, a shared experience, a common structure, a joint territory, a ladder of communication; and together we climb it. Such a ladder requires equivalent sharing of the commitments, transparencies, and relation that make us who we are. Each advance in shared context permits greater information exchange, which in turn permits the accumulation of additional context. To make this work of course requires also a commitment to veracity in the relationship; otherwise, the communication is untrue, the context is built with flawed materials, the trust is betrayed, the ladder falls.

I have put this sharing in the language of information theory, because of its applicability to any form of intelligent life and ultimately to questions of art and civilization. An equivalent interpretation of experience may be found in the study of psychology (about which I admit to understanding less, except for knowing a good many psychologists up pretty close.) Here is Carl Rogers, for example:

> Assuming (a) a minimal willingness on the part of two people to be in contact; (b) an ability and minimal willingness on the part of each to receive communication from the other; and (c) assuming the contact to continue over a period of time; then the following relationship is hypothesized to hold true[:]
>
> The greater the congruence of experience, awareness and communication on the part of one

individual, the more the ensuing relationship will involve: a tendency toward more mutually accurate understanding of the communications; improved psychological adjustment and functioning in both parties; mutual satisfaction in the relationship. [2]

When this letter started writing, it began wrongly: "If experience is a teacher, this family's love has made me an expert." Such love, however, is the kind that rolls easily downhill from great reservoirs, incautious and profligate, without effort. It is greatly to be desired, but it is special and private. One should seek teaching from the love that is hard to come by, that needs some pumping uphill, rising with care and work, but more universally available. Those who have found their love among the pained and incomplete have such teachers.

In our universe, things tend to flow downhill because it is the easiest direction and therefore much more probable. We live in a probabilistic universe. Water flows only downhill in everyday experience, and so does energy; in the latter case, the gravity gradient is replaced by a temperature gradient (that is, we observe that heat tends to move from higher- to lower-temperature regions if unimpeded.) It is not impossible for a molecule of water or a photon of energy to move against their respective gradients, but the crowd is so large that their average sentiment always prevails in our observation.

One can, however, pump water uphill and one can pump energy against a temperature gradient (witness the refrigerator.) This requires the expenditure of work, some of which is not recoverable because few such processes are fully reversible.

Love also tends to flow downhill. We most often experience love, and tend to seek it primarily, from particular, closest sources. The benefits and gratifications we associate with love tend to flow spontaneously "downhill"; i.e., from those with whom a context has been established and attenuation minimized, who are already engaged and inclined to supply them.

But we can also pump love uphill from a larger number of non-engaged sources, just as we can extract heat from a larger number of less energetic particles (i.e., at lower temperatures.) It just takes some work. (Engineers consider "work" to be that form of energy into which intention has been injected; it differs from random activity by its direction or coherence.) Our problem is that we are conditioned to believe that pumping something as important as love uphill is beyond our capacity, that love "happens," can flow only from particular sources in sufficient quality or quantity to be fully gratifying.

The universe, however, says differently. By doing some work, one can pump the most valuable things uphill. One can pump good winter heat out of a cold lake, and summer heat out of a refrigerator, and a clear message out of louder noise. It is a matter of collecting and orienting lots of little pieces together, finding the coherence under random processes, getting particles lined up in the same direction, integrating long and wide receptions until the buried signal rises above the surrounding noise. It is true: the universe permits this. Organic life evolves in just this way. There is no reason why love should not behave the same.

Those who are surrounded by supportive friends and family are the beneficiaries of love's slope. It flows down to them with little effort on their part. This does not make them wise, and they need the example of those who have learned to perceive the same benefits in relationships that require work.

So we need to strive for connection (Buber would say relation) in each context, not wait for it; that is a lesson in all the places from which I write to you. How do we sometimes find it? Where do we miss opportunities? Why do we fear to touch? What boundaries are we even now erecting, or imagining others to erect? Some insights on these questions await us on a little island off the coast of Maine.

Notes

[1] Martin Buber, *I and Thou*, trans. Walter Kaufmann (New York: Scribner, 1970), 57.

[2] Carl R. Rogers, *On Becoming a Person* (Boston: Houghton Mifflin, 1961), 344.

Monhegan Island

ZERO MOSTEL GREETS every boat, it seems, and he is heard long before he can be seen. He is unsubdued by this quiet island, uninhibited on the dock and paths, on stage. He carries his stage on his back, pushing his presence ahead and drawing others behind, through the little village and along the lobster traps neatly stacked for another season, past the artists trying to capture the scene in peace, up the hill to the softball field just past the lighthouse. There he pitches, of course, and there is my only claim to an acquaintance.

The every-Sunday game draws artists and businessmen, lobstermen and technicians, students and teachers, all feeling far from their respective pressures and engaged in a different competition. There is more laughter than athletics to enjoy, and Zero's antics prevent any of us from taking the game too seriously. He makes me think about the importance of humor in our culture.

In our cherishing the capacities of life, humor deserves a higher place. I should expect to find it in other life forms that we may encounter around the universe. It rises from the same base as our perception of beauty and our search for knowledge, thrusting upon us a burst of reality, a high-density dose of correlation.

In its better moments, humor confronts us with a sudden connection, a startling communication of common experience seen in a new way, a sharp evocation of an incompletely-realized truth, a new and creative linkage, a shared surprise. Even Zero's least sophisticated jokes and antics offer something novel, unexpected, out of place, inappropriate—sometimes empathic, sometimes grotesque. Humor is important because it communicates, and does so with engaging intensity.

Zero's humor on the dock and ball field is not always sophisticated, often crude, but he has captured on stage and film its beautiful dark side. The dark humor we learn from the incisive observer in poor and oppressed circumstances is funny and comforting for the same reason: it facilitates sharing.

This does not need great pain to appreciate. Everyday frustrations and limits intrinsic to the human condition give us the context to use humor to our comfort. Humor is a safe way to reach out; we are threatened by close proximity and sudden approaches but can risk a bumper sticker or a joke. When we are delayed we can quote Hofstadter's Law to each other: It always takes longer than you expect, even when you take into account Hofstadter's Law. [1]

Humor builds bridges, brings us together, advances our progress toward a whole people. It is integrative, and that is what makes symphonies out of notes—or, if you prefer, machines out of parts. One escapes limitation by the freedom of connection. Humor frees us—for a moment—to risk connection, to try it out. Occasionally we escape.

* * *

UNHAPPILY THE REACH of humor is more often brief and tentative. Its very suddenness, its engaging intensity, the thing that makes it work, makes it perishable. It goes stale right after it gets fresh. It lacks the courage to dare uncertainty for a long engagement. In fact we often use humor to cover a retreat, to avoid a serious encounter of any duration. Beneath the bluster, Zero wants us to love him, but we sustain only the banter.

My mother, who was no Zero Mostel, carried with her a joke from the nineteenth century, and shared it with me:

Girl: Nobody loves me, and my hands are cold.

Boy: God loves you, and you can sit on your hands.

Beneath the joke lies a deeper pain and colder comfort. God is not an accessible person for many of us, and we personalize love, placing its source in one or a few individuals. We seek love in a particular person, and it seems gone when the person leaves. Surrounded by others, we feel one step from alone.

Humor allows a fleeting touch. But it is not dependably beneficial, as easily demonstrated by reference to many jokes based on stereotypes, and can be cruel and exclusive. We need a longer embrace, and that comes only with vulnerability. Shannon tells us that the receiver must open its aperture and welcome the unknown if it is to obtain any information—to derive any value—from the transmitter. [2] So we have a form of introduction, an approach. Our need is to sustain the communication and to make it inclusive.

* * *

WELL, MONHEGAN IS a peaceful place, and we have come to rest. We do not always seek or welcome new encounters. Our banter is like the fog's advance and retreat, cold and transient, without lasting commitment. I wonder at this as your mom and I watch a wind-chased fog race through the fir forest, so unlike Sandburg's little cat feet, and I write her a poem:

On this island fog does not creep in
> but strides and plunges through the wood
> > as from a great, cold fire.

A brief embrace of fog, just long enough
> to leave a gift, a kiss of moisture
> > from the sea and stone,

A chilling sea-stone's kiss, leaves only this -

a greeting or a parting gift?
It will not say: the coming

And the leaving's all the same with fog.
One cannot know in brief embrace
what parting may already

Wait behind the tenuous touch, the tender
but unlasting hold. So let us
dare uncertainty in

Fog condenséd of our fears. If love
allow a clearing of such fog
for two, then clearing may be
made among us all.

Some say there are ancient writings on the Monhegan rocks, and your mom and I have looked for them. We know that somewhere near here humankind completed its first encircling of earth, the Norsemen encountering Native Americans and then withdrawing. So these rocks may have witnessed our first encounter with ourselves, and many since, each tentative touch and retreat mocked by Zero's humor and the indifferent wash of fog. (The next introduction of Europeans to Native Americans would be more lasting but not more benign.) [3]

And so we continue, advancing upon one another and retreating, seeking more advantage than cooperation, responding to difference with hostility. There is a kind of fog we make among ourselves, exhaled about us like a cloak, sparing us from connection, saving from commitment. Through the fog and from the carvings on these rocks, I try to read a lesson about what might have made these encounters succeed.

How shall we sustain the engagement? Our religions urge us to love one another, generally sound advice but hard to follow while romanticizing and particularizing the concept of love. We look around at the other visitors on the rocks and the boat: can we love them all, not to mention the unappealing millions at greater distances? Entering a "love" relationship with many in such a crowd seems beyond reach for most of us in gentle times. And for those threatened by invasion or competition or attack—real or imagined— the aversion witnessed by these rocks over the centuries is predictable. Loving one another seems hopelessly impractical as a guide for real living.

The difficulty is not limited to our transitive reach. We despair also of deriving comfort or identity from the many with whom we are not already closely engaged through long and intimate sharing. Losing a "loved one," in physical or emotional separation, seems impossible of replacement, irrespective of millions who may surround us. In stress it is hard to be confident of our ability to "pump love uphill" from less engaged sources, as we advised in Shaftsbury.

So we are inhibited, afraid, pessimistic of either giving or receiving love to or from the many—because we cling to a narrowly focused, exclusive, irrational sense of love. Yet if we are right in our rational understanding of love, both its demands and its extensions should be within realistic reach. Substitute for "love" the concepts of commitment, trust, and extension, and the possibility of their building, step by step, appears more likely. Like life, love in this rational sense is expressed in degrees and allows us the potential to grow. In fact we can make more and stronger commitments to others, and we can learn to trust increasingly, through our capacity to communicate.

* * *

CONSIDER THIS: the genetic variation among us is estimated, as I write, to be less than one out of a thousand bases in the sequence of nucleotides that make up the structure of our DNA. We are 99.9%

identical genetically. That first Norseman to encounter a Native American tribesman came into the world different from him by only one tenth of one percent. We have the same six senses,* the same morphology, the same means of moving about. The media of our communications are shared, as are, to a large degree, our concepts of beauty and morality and our understanding of physical reality. Even our traditions and religions have more in common than in real distinction. We have evolved to be acutely aware of differences— however tiny compared to the similarities that fill in the vast back-ground, that give us psychological "construct," that provide context in our communications.

So when we look at another we are looking very much at our-selves. This is true. Perceiving this fact frees us from the lock-focus on difference, allows the exploration of wider sharing. It is a stronger concept, and one that allows us to derive that extension of self we seek, not just from one but from many.

* * *

COMMUNICATION IS THE medium of love, as of life, the medi-ator of trust and commitment, of connection and caring. Seeking the broader Context and including Veracity and Acceptance in our communicating [2] reaches beyond humor's tentative touch. Love is the trust of surprise. It fills in the context in communication, which permits and supports sharing and daring. The wider concept of "love" is a natural and rational expression of intelligent life, a com-mitment and trust and mutual extension that must be experienced in all advancing life forms for them to prosper long. The unexplored potential of sharing enables finding a wider identity through others.

It is within our capacity to reach beyond close but insular relationships, to escape the romanticized and passive notion of

---

* I count the kinetic sense—our ability to detect accelerations—as number six.

love as something that happens to us, and to grasp the rational understanding of love as a capacity that can be nurtured. Try it.

Notes

[1] Douglas R. *Hofstadter, Gödel, Escher, Bach: An Eternal Golden Braid* (New York: Vintage Books/Random House, 1989), 152.

[2] See the letter from the RCC laboratory, esp. p. 33–34.

[3] See the letter from Hawikkuh.

The Metropolitan
State Hospital

THIS IS THE most frightening place of all. There are no
weapons here, no fences, no fights. The very absence of
contention is the institution's most chilling aspect. No
one would choose to remain long in this gloom, haunted by things
unreal, yet the residents stay on. Most could walk away at any time
but do not. Like Poe's novels, the subject of the place is not violence
but madness, not external but internal, and we risk being drawn in.

Neither hospital nor prison corridors convey the sense of
hopelessness that pervades this large institution for the mentally ill.
Not all the patients are permanent residents of Met State, but nearly
all are incarcerated by closer walls than those surrounding the prison
and the wards. Their walls are within them, and I know they could
be raised quickly within me, and that our treatments are still long
and unsure. It is far more frightening to walk these halls and the
grounds outside—beautiful grounds on some of the best real estate
in suburban Boston—than to move around the prison, because so
little separates our outside from their inside.

What makes that separation? What have we on the outside
experienced that they on the inside missed? Many of the patients
here suffer from congenital or imposed misfortunes, but most missed
something early in their lives that is very hard to restore later.
Without it, in fact, children in general do not prosper and adults
do not mature. For those whose early experience misses any mutual

commitment and trust and extension of self, the capacity to communicate and to relate—essential elements of living and being—grows
with difficulty. We see the results every day in those assigned to the
Departments of Correction and of Mental Health.

Dr. Adams,[N] who works with both populations, tells me about
the different adaptive and defensive strategies developed by children
of chaotic and uncaring homes, explaining in clinical terms why
some end up on the streets, some at the Middletown Prison, and
some here. As my hours in all three places accumulate, the clinical
determinants are overwhelmed—for me as for them—by the heavy
realities of their present states. Can we understand this experience
that so affects us, the deprivation of which might cast us into this
frightening inside world?

Buber would call it "relation." It is, by any name, an indispensable part of who we are, the part that supports extension of our
"selves." It is the long reach asked of the members at the club, the
absence that makes men "numbers" at the prison, that for which the
Newfern back ward cries, the final freedom. Here at Met State we
see most starkly the effect of its missing.

Such an absence can cause any of us to raise walls within
ourselves, defending a fragile psyche against further pain, disabling
our interactions. At Met State the walls are so clearly inside, instead
of outside, that we come away frightened by how little separates us.
In fact we are all mentally ill from time to time and the distinction
between visitors and patients is a matter of degree.

* * *

THIS IS OUR greatest fear, the despair of nonbeing that has stalked
all our philosophies, as traced by Tillich. [1] Intelligent life depends
on freedom and the choice of order. What if both could be taken
away—what would be left of us? This happens at Met State, and we
are forced to look at it. None of us is in full control at all times, but

we can disregard that reality and go through most of our days with a generally secure sense of self. Each of us feels he can command his own behavior, determine her own thought; each feels confident enough to plot a course and affect future events:

> I am the master of my fate,
> I am the captain of my soul. [2]

Not so. Not at Met State. Here we must recognize that the familiar "self" can be destroyed by accident or manipulated by psychological, chemical or electrical forces quite beyond our control. We walk down the hill from this dark encounter quite changed.

From now on our recourse must be to each other. The "individual"—not sharply defined anyway—varies in both quantitative expression of life and qualitative expression of being, of self, of identity. Each of us can make commitments, to a degree, and be transparent to one's values, most of the time. To be faithful is to do one's best within such limits. One cannot consistently command, however, either external events or internal direction. As in facing death, we must find our selves in relation.

Such finding is supported, I am persuaded, by the reality of intelligent life, based on capacities that are shared. Our freedom and ability to organize, to build, to choose order, to see and make beauty, to learn, to heal, to love, and to communicate are the characteristics of intelligent life generally, not capacities fully expressed in the isolation of individuality.

Heinz Pagels quotes Einstein:

> A human being is part of the whole, called by us "Universe": a part limited in time and space. He experiences himself, his thoughts and feelings as something separated from the rest—a kind of optical delusion of his consciousness. This delusion is a kind of prison for us, restricting us to our personal desires

and to affection for a few persons nearest us. Our task must be to free ourselves from this prison by widening the whole circle of compassion to embrace all living creatures and the whole nature of its beauty. Nobody is able to achieve this completely but the striving for such achievement is, in itself, a part of the liberation and a foundation for inner security. [3]

This quotation captures for me the psychological and philosophical essence of Met State and its projection on us all. The delusion of which Einstein speaks is indeed a kind of prison for us, and the restriction is greater only in degree for those inside. We all suffer from the experience of separation, some more than others, and it is always disabling. But Einstein points out that we can be liberated by widening our circle of compassion.

A larger sense of identity is necessary to avoid isolation, the walls of Met State. We have considered a dependable identity to be an expression of commitment, transparency, and relation. I believe Einstein, Dr. Adams, and Martin Buber (who gave me the term) would agree that "relation" is the *integration of one's extensions*. By this I mean that the element of one's identity derived from "relation" grows as a function of one's experiences with self-extension.

Each communication with another—whether direct or intermediary, whether auditory or visual or tactile, via beauty or building, learning or healing—each communication beyond our narrowest self contributes to a store of experience that I have called the "extension" of one's self to others. Some of that experience is positive and some negative, broadening or narrowing one's sense of self. A close and faithful experience of love involves thousands of such communications, stretching one's being toward the other. A destructive or unfaithful experience tends to shrink one's over-all "extension."

The integration of this store of experience constitutes what I have called "relation" in the expression of one's identity. (I use "integration" here in its mathematical sense, to take into account positive and negative experiences and extensions relative to a variety of others.) Thus we grow, we "become," we approach a liberation and a wholeness over the course of our experiences with others, mediated via communication in all its forms. This I take as consistent with Buber's meaning of "relation" and with Einstein's "striving for such...liberation."

* * *

THIS YEAR I AM chairman of the Governor's Council on Mental Health and Retardation, formed along with area boards and regional councils to give a citizens' voice to the sweeping process of "deinstitutionalizing" our MH & MR patients. In a few years this will be largely—although not altogether effectively—done: the thousands at Met State will dwindle to hundreds, then scores. In the meantime, we are engaged mostly in legislative and budgetary issues, and often forget the real pain of patients and their families.

Our meeting room is in a cavernous facility in the Government Center of Boston, built to house both treatment and administrative functions, combined to remind us of what we too easily forget. Sometimes the combining works: last week I shared the elevator with a mother and her child who were seeking treatment that I know is still unsure. The child was crying on my elevator, frightened, and I was suddenly a parent instead of a chairman. I wanted to embrace them, to help, to speak assurance, but I had not the skill nor the confidence nor the right. I could only resolve to carry the encounter with me as I steered our meetings and sought more resources.

The "relation" expression of our identities, lost or never developed by the patients in our elevators and institutions (and imperfectly in all of us), can be integrated and nurtured only with the help of others. All our positive experiences of self-extension—mutual trust and commitment, healing and defending, building and learning, sharing of beauty, acceptance and forgiveness—all such experiences rely on those in communication with us. We shake off Einstein's delusion and find liberation from our own prisons not by heroic escape but by drawing on others.

Rebuilding this capacity for relation is what Dr. Adams says he and his staff spend all their time on. Being able to trust, to make commitments, and to extend oneself to others is a learned skill, he says, like language. It requires practice in safe places, time to learn and to try, examples and reinforcement, gradual encouragement of one's reach and the slow, careful building of the courage to be vulnerable.

The lessons of Dr. Adams, Einstein, and Buber are consistent with our understanding of information theory. An expanding shared context supports trust and therefore permits a tolerance of uncertainty, opens one's aperture to a larger content. This process continues while it meets the test of veracity, further reinforcing trust. It is a kind of feedback loop in the expression of one's identity. Communication supports extensions which integrate in relation, and the "becoming" of relation facilitates fuller communication.

One may propose exceptions to this process—parenting, for example, or romantic or sexual relationships—intense and appearing to obviate the slow building of context, trust, and commitment, being more immediate and organic. But these apparent exceptions are based on contexts that are given or assumed in advance, a head start up the communication ladder. We are willing to stipulate trust and commitment: the infant trusts implicitly in the parent, the parent (usually) makes an immediate commitment to the infant. The same is true, or seems to be true, in romantic and physically-intensified love. These commitments and trusts do not necessarily last, however; the instinctive, given relationship needs to be supplemented with a more mature, communication-based bond, a dependable context, daring vulnerability, proving veracity, not relying on organic or irrational precepts.

* * *

FOR THOSE AT Met State, there is nothing to supplement: earlier attempts at extension were unfulfilling or destructive. But all of us have experienced pain—often our greatest pain—in the failure of relation and its impact on our own identities. All need to experience mutual commitment. If love were perfect, the commitment would be unconditional, like the prevailing of truth in every event. This gives us trust, security, a constancy amid change; it allows us to be open to change, because we have a base, a grounding. Because few commitments are unconditional, however, the "relation" element of identity depends on a gradual integration of our more successful extensions.

Love supports communication and perception. In a committed, trusting relationship, one can afford to be more vulnerable, more tolerant of uncertainty. We exchange information in proportion to the uncertainty that we can deal with. I receive better because I risk the unknown. I transmit better because I risk exposure. In a widened identity, in which relation is a strong element of self, I can communicate, as it were, with myself; thus the distortion and attenuation of an artificial channel dissolve. This is extension. We become transparent for each other, to what is real. The trust relationship means we expect to see truth in the communication.

So love allows us to widen who we are, to extend ourselves through the relationship, to be more whole. But what of "caring"? We are not entirely confident of another's commitment, extension, or trust—these rational explanations of love—without evidence that he cares deeply for me, that she cherishes my company, identifies with my well-being, more perhaps than with her own.

Exactly. The closest bonds are made of extensions in each other. I experience love most deeply when the other seems part of me. If she were gone, there would seem a hole in my being. My wish to be close, my pain when she is in pain, our special sharing, all grow with the growth of our mutual extension. It is because she is part of me that I hurt when she suffers or is gone. Our strongest feelings always come from tearings in the fabric of our identities. There could be no stronger evidence of caring. The love experience is an expansion of self.

This capacity to extend to each other, widening our "circle of compassion," is the becoming we seek. Evolution is not completed in human development but accelerates as a people finds better means of communicating, higher order in building together, greater beauty in mutually-supported discovery, expanded freedom in collaboration—i.e., as a people finds its shared life.

Notes

[1] Paul Tillich, *The Courage To Be* (Yale University Press, 1952.)

[2] William Ernest Henley, "Invictus" (1875.)

[3] Heinz R. Pagels, *Perfect Symmetry: The Search for the Beginning of Time* (Simon and Schuster, 1985), 362. (Pagels does not give a source for this quotation of Einstein, but others have attributed it to him also. In any event, Pagels himself would be a sufficient source for me.)

Dorchester, Massachusetts

I T IS HARD TO imagine a farm in this paved and crowded neighborhood of Boston, in which little of plant or person grows without effort. Yet here is a little park, squeezed among three-deckers and neglected apartments, littered with the plastic scraps and wastes of a throw-away urban society, a park that was once the highest field of a large green farm. Its bordering streets are noisy with machines, its air dense with toxins, where once the sounds and fragrance of plowing and harvesting prevailed. The center of Boston, now a short subway trip away, was once remote, reached by carriage in expense of a good piece of morning.

The Boston churches were remote enough to prompt Abigail Adams Eliot's grandparents on the May side, one Sunday morning a generation after the American Civil War, to take the shorter trip from this farm to the First Parish of Dorchester, to hear a young minister named Eliot. His impact was substantial, especially on the Mays' young daughter, who eventually decided to marry and bear him three children, of whom Abby was the third. Abby's brother Frederick May Eliot would become an especially revered president of the American Unitarian Association; her sister Martha would receive one of the earliest MDs granted to a woman, do some pioneering work on the eradication of rickets, and become chief of the U.S. Children's Bureau and prominent in the World Health Organization; and her cousin Tom would move to England and write some of the great poetry of the twentieth century.

Abigail Adams Eliot herself would live a hundred years, and fit into that space three careers: in child care, mental health, and elderly services. She is widely considered a pioneer in at least the first two, and the world is a different place because of her work.

I have brought Abby here to Dorchester to see her grandfather's old house (now multi-family) and the park that is left from his fields, and the First Parish with her father's name carved in the entrance. She cannot see much but I am her eyes, and she is mine. Through her I see back to the turn of the century, listen for the carriage sent rattling over the Beacon Hill cobblestones to fetch her and her sister to Dorchester, hear the speech given through a megaphone by Edward Everett Hale from the State House as the twentieth century dawned. The Rev. Mr. Eliot wanted his daughters to be in that crowd, and woke the little girls just before midnight to walk up the hill from West Cedar Street. [1]

* * *

WHAT MIGHT EACH of us do in a hundred years? What means are available to us, if we are to be faithful? The meaning of life was not a mystery to Abby: it lay in an unconditional commitment to truth and the benefit of life. I cannot think of a better place to look.

It is, of course, how we *use* the capacities of life that establishes our identities. Our commitment to the benefit of life might then reach to expanding the identities of others, to helping them with meaning as well as living. This is what Abby did for us. Her means included assisting others with their commitments and relation, which constitute the moral dimension of intelligent life. Logically if we have found our meaning in a commitment to truth and the benefit of life, we would wish others to find the same. The benefit of life requires its fulfillment as well as its preservation, its being as well as its living; and always its sharing.

Then I think Abby would allow me to add a third element to her lesson on the meaning of life, one I would call "becoming." In the sharing and honesty, the helping and searching, that she exemplified, there was evident a continuous growth. Her identity was not static, nor was her vision of what we could be. In her pioneering work with young children, as in her later guidance of mental health programs, there was a faith in our ability to be better, to mature, to grow through mutual effort. This seems to me a trend of evolution and civilization, a promise and a hope—that we can mature as a people, that our "becoming" is incomplete. Abby believed she could make a difference, and so do I. Our impacts are multiplied through the engagement of others, and we become together. The geometric progression of our understanding of truth and our ability to help each other proceeds from this multiplication.

These opportunities, as a guide to fuller living, derive from a rational view of the universe. They are not dependent on religious dogma and require no affiliation with an institutional church or prescribed creed. If we start with what we can comprehend, make a commitment to truth, and seek to expand both our understanding and our connection, they obtain logically. This is not to say they are inconsistent with church affiliation or with the faithful pursuit of any religion, only that their base is in truth rather than in human invention.

* * *

WHAT MIGHT WE call a hundred years of such committed living? Truthful and faithful, surely, but something else: I call it ablative.

In the thicket of technical problems that had to be solved early in the U.S. space program, one vexing vine was named "re-entry." How might one bring back, preferably unscathed, the astronaut launched against earth's gravitational field, when that field had been observed to accelerate meteors to such a velocity as to vaporize most of them totally by friction with the atmosphere, long before they reached the surface? The solution adopted for the earliest flights was an ablative shield, which would absorb the heat of re-entry by gradually vaporizing. Such a shield would seem to be mostly gone when its purpose was accomplished; it would be shrunken, rough, scarred, misshapen, no longer functional; but its mission would have been faithfully served. The first such shield was made simply of fiberglass strands and phenolic resin, and it worked. Later ablative materials were more complex but the first had been sufficient.

The components of an ablative thing tell us little of its purpose. When they are gone the purpose is served; their whole is realized as their pieces are disassociated.

Abby, it seems to me, lived an ablative life. Such people give away their substance, in the process becoming transparent to their commitments. In the end that is all that may be seen, and it is more evident to us for their transparency. A faithful parting of the opaque walls of our divisions gives special clarity to the possibilities beyond.

In striving to "be," suppose we were to take a radical approach and give away all, even our selves. Might more in fact remain than if we struggled to retain our selves? We could give away what we build, to attach a dependable purpose to building. We could give our resources to the increase of freedom and connection, to enhance those capacities of life among those who share it with us. We could finally give up our physical bodies, as a gift to those who follow.

Abby lived an ablative life. She gave away everything she had, went on giving for a hundred years. In her eighties and nineties, much of my time with her was spent helping her to give. She taught me this:

* * *

YOUR CHILDREN WILL eradicate disease, secure comfort, find new pleasures. For them the challenge is to bring together those who contend, who are isolated, who bear old grudges, who take offense.

We need the people in pain, Abby said. We need those who trouble us, who bring us discomfort, who make us worry. To find harmony we must confront discord. We cannot experience the best of love by limiting our relationships to those that are gratifying. In our struggle to become whole, we need the help of those who need us. Without them, we are denied the opportunity of reaching out beyond the nearest horizon.

More than that, we need those who are hard to help—the ungrateful, surly, unattractive, dishonest, threatening, resistant needy. Every unpleasant attribute we can perceive in others is present within ourselves. We have our times of being ignorant, irrational, unethical, inconsiderate, dirty and unpleasant. If we cannot reach through such attributes to others, how shall we be in touch with our whole selves?

Is this too great an assignment? Your children should remember, Abby said, that they do not have to be wise or strong in every event, that it is not their responsibility to make others happy. Their daily task is to be as transparent as possible to the truth that is already present, and as opaque as possible to falsehood. It is to strive for a clearer discernment of one from the other, and of the presence of truth in others and themselves. It is to see more of what is real.

They need not work miracles: they are already worked. Truth is present everywhere, and the potential for life, and with it the possibilities of love. Your children need not make others happy, or provide their love or define their truth; indeed they cannot reliably be the source of good, but they can be a window to it. If I am the maker, the provider, the definer, then when I am gone where will be those whose nurture I have provided? But if they may see through me, then their own vision is extended, and they are nurtured by what is always there.

* * *

I REMEMBER ABBY taking charge of a room full of sophisticated professionals, speaking out spontaneously, organizing us into committees and assigning tasks, by what seemed at the time the force of her personality and intellect. Years later I could see that the force came from her obvious commitment, from her caring for us and ability to relate to us, from the courage and perceptions we could see through her. Now I am walking slowly back to the visitors' parking lot from the nursing home that she will not leave again. Is that the same Abby, shorn now of memory and dignity, turning her back to me? Only a little bit, I think; some days a bit more than others.

Then where is she? Is such a force eradicated by time? As I walk slowly to my car, a response forms in my mind: Abby is in here, in me, at least in part, and in the others she loved and helped and taught. We are Abby now.

Abby was for us a window, a gift. For us she was transparent to truth. She lived an ablative life, investing it in us. When I say she will always be with me, I am making neither an emotional nor a mystical claim. It is rational, and because of that (especially for Abby, who would not tolerate foolishness), it is faithful.

While Abby was waiting for death, we spoke often of life and its meaning. I told her that she had given her life to me, and to the thousands of others whom she had nurtured and taught and organized and inspired over the long years, and that she would live in me and in them. She liked the thought, and held it.

* * *

So THAT IS Abby's lesson for you, Mo. She lives in me, and now in you.

Notes

[1] Abby told me a hundred stories over the last two decades of
her life, and dictated many on my machine to make a book,
ably transcribed and edited by her friend and associate Marjorie
Gott Manning: Abigail A. Eliot, *A Heart of Grateful Trust,* ed.
Marjorie Gott Manning (Medford, Massachusetts: The Eliot
School of Tufts, 1982.)

Hawikkuh, Zuni Land

AWIKKUH IS NOT much of a promontory as mesas go. Its place is obscure, archaic on most Anglo maps, but its view still sweeps the desert plain below. Our unpracticed eye scans from the drop at our feet to the far horizon, to the east, the south, the west, no direction distinguished from the others, only random mesquite and the rubble of wind and time. The southern view is why these ancient stones were laid here, but they had as much domestic as military purpose, including the holding of the winter sun on this high plateau.

On this August afternoon the sun seems less a friend. The Zuni River that made the valley long ago is nearly dry. No evidence of moisture relieves our anxious gaze, except infrequent scrubby growth and the far-off dust of sheep herds, both reaching toward sources of water that we cannot discern. The empty air has let these walls remain a few stones high, centuries after the Zuni moved north. It is a purifying air, thin and unredeeming as we climb, without fragrance, desiccating, sterile except in movement across the desert floor.

The pace of time cannot sustain itself in such an air and slows, out of breath, to wait the seasons. Its elusive arrow lengthens and almost bends for us, taking direction from more gradual disordering. For this is the nature of time: that neither its pace nor its direction, forward or back, are intrinsic, but given by the proclivity of our universe toward disorder. It is nature's downward slope that gives us time.

One can read nature's slope in the stones of Hawikkuh. They are rough to touch, not yet defeated by gravity and wind, and the mortar rougher still. I move my touch along its lines as the builders did, my fingers reaching for theirs. Some of the grooves still fit fingertips, and time, the only space between us, dissolves in the caress of stone and the embrace of wind.

* * *

THE GROUND IS mostly sand and stone made of sand, evidencing the waters that once stood above this plain. It is red-brown on the whole, local variations from nearly-white to dark-gray not apparent at any distance. It has a definite texture, even at a distance: rough, sandy, ablative, stratified, linking our senses of sight with touch.

The people themselves share the color of the sand and stone, and they believe in a deeper sharing. The Zuni have their own language, unlike other tribes who share common roots. To me, this gives them a special continuity in the mystery of time, in which an ancient reverence for the living earth prevails. So their open temple is a good place to bring questions of order and time and life.

> "What geomancy," asks Annie Dillard, "reads what the wind-blown sand writes on the desert rock? I read there that all things live by a generous power and dance to a mighty tune; or I read there that all things are scattered and hurled, that our every arabesque and grand jeté is a frantic variation on our one free fall." [1]

The generous power and mighty tune have been the objects of our hopes and worship since the Zuni ancestors followed herds from Siberia, since stone-age peoples raised circles of monoliths on the English plains, since we began to paint and dance and sing. We have tried to live by the generous Power and dance to the mighty

Tune, and many still do; but as mystery recedes we feel scattered and hurled, and read in the rocks our one free fall down nature's slope: the inexorable decay of order in the universe, the inevitable growth of entropy.

I read, however, something else in the Zuni sandstone. It is not a supernatural carving but a beautifully natural capacity of intelligent life to recognize order and to choose it: to raise stone upon stone in freedom and joy, to seek and to wonder, to find beauty and make it, to reach out, to learn, to heal, to love. Is this not worship?

I will not choose between dancing to a mighty tune and falling frantically into chaos. Something there is that is both faithful and rational, both beautiful and true. We can compose a mighty tune; can we transcend our own boundaries and find a generous power? I do not find limits to our purchase of order, to our reach.

* * *

ALTHOUGH THE HAWIKKUH walls have tumbled, their view of the plain below has not changed much since 1539: here the native people of what we call New Mexico first watched European visitors advancing upon their land. Across the Zuni plain came the Spanish Moor Esteban, with a group of native people from the southern land we call Mexico. They were an advance scouting party for Spaniards seeking gold and converts to Christianity, and Hawikkuh was thought to be the first outpost of Seven Golden Cities of Cibola.

Somehow Esteban offended his hosts, and his career was abruptly terminated; his employer Coronado, however, was not easily dissuaded, and the flood of Europeans could not be diked. They returned the following summer, and in all the pueblos today the Catholic churches stand near where the kivas lie in ruin. Gold, on the other hand, was in shorter supply than converts. Ultimately all the Westerners who would invade this sparse land would leave it to the people who had struggled over the Siberian land bridge during the last Ice Age to claim it. Unusual among "reservations," this is the primeval home of its residents.

Mostly quiet prevails today. There is always the wind, and songbirds in the warm months. A flock of sheep graze and search below. Occasionally the offense of military planes breaks the quiet, and we try to imagine what Esteban's view of the future would have been.

Hawikkuh was partially excavated around 1920 by Frederick Webb Hodge of the Museum of the American Indian. He reached back only to the Spanish roots, however, and left a mess. Further excavation to reach the original inhabitants remains to be done. Thousands of potsherds lie even on the surface; since the place is not easily found by non-Zunis, the wind has been the principal exhumer. Remnants of walls, depressions, mounds, an occasional cairn, mark the places of promise. The apparent village measures about a hundred by two hundred meters, not counting the mission church built in 1630 on lower ground to the southeast (and burned three times by the Zuni.)

* * *

WE LEAVE HAWIKKUH and drive a few miles across the desert to the east, to seek the more ancient site of Kechipauan, inhabited in the eighth and ninth centuries. Some cairns mark what may be its place, and we begin looking for confirmation in the stones. What are the signs of intelligent life that will satisfy a rational search? A pile of rocks here, a suspicious rise over there, a straight row, a circle—we want to find a pattern that someone like us might have made a long time ago. It is the same process pursued by the radio telescope on Harvard's hill: in each search the same watch for that great improbability, for intelligent life.

There are some intrinsically improbable arrangements: a sustained vertical arrangement in a gravitational field, for instance, so a free-standing wall qualifies. (Well, if it's high enough; the higher the wall, the less probable a non-intelligent cause.) But a partial circle of stones or a short straight line do not qualify; they are merely familiar, no less probable than what we consider a chaotic jumble. (In fact, there are many "natural," non-intelligent causes of straight lines, circles, hexagons, and other symmetrical arrangements in our universe.) Any particular jumble is as immensely improbable as any familiar pattern, and maybe more so; we just don't recognize its particularity. Only a continuation or proliferation of such familiar patterns will accumulate satisfactory evidence, and then only because we have the capacity to recognize something that we think another person also recognized. We can judge the probability of a human cause to be greater than other explanations because we know in advance what sorts of things humans build. It is the accumulation of familiarity that convinces us that someone has been here: the higher the wall, the rounder and larger the circle, the more likely its intentional construction. Still it is not certain, but as the height or roundness increases, our belief strengthens.

Symmetry will not do, nor coherence; the universe generates plenty of both spontaneously, without apparent purpose. Only when we *share* our perceptions of "order" does it take on the quality of intelligent intent. This perception depends on an *a priori* frame of reference; when I speak of perceived order, I am appealing to the sharing of that reference frame, that context. Thus I am talking, at root, about communication. We know something about the people that may have placed the stones, and we *correlate* our observation with what we know. The probability of that people's placing the observed pattern can then be judged higher than the probability of non-conscious causes of the same result.

It is the *context* of such patterns that makes them seem intrinsically orderly: because we have a shared experience (the use of circles or straight lines), there is a context established in advance. Random events are as likely to produce a circular or linear pattern as any other *particular* pattern, each one of which is unimaginably improbable. The difference is that we have conspired to define a special set of patterns and to sustain the definition long enough to admire them together.

* * *

WE RECOGNIZE ORDER as an indicator of life. We build cathedrals of one sort or another, stone on stone, ceremony piled on tradition, regulation upon law, religion on hope, seeking unrecognized order and hidden connections in the wealth of possibilities presented by an uncertain future.

To the early Christians in Jerusalem, an anonymous writer sent a long argument, probably before A.D. 70, to keep the faith. He began its eleventh chapter by defining faith as "...the substance of things hoped for, the evidence of things not seen." [2] What is it we hope for on Harvard's hill or on the Zuni mesa? What is the substance, the evidence, of the intelligence for which we reach, and hope? How do we recognize the capacity for order, the thing unseen, in the midst of disorder?

Last night we watched the sun set between twin mesas to the west of Zuni Pueblo. The beauty we perceived lay in our minds' ability to recognize order on a grand scale of color and pattern. We associate beauty with a shared recognition of an unlikely but recognizable (communicable) pattern. Beauty is a celebration, a stretching, of the skills of consciousness. We have the capacity to discern order, to correlate, to anticipate and remember, to wonder— these fundamental skills of conscious life are exercised by our art and our perceptions of beauty. It is not "order" in itself, but the sharing of its perception that underlies our art and our civilization.

To reach and maintain any *intended* state when a large number of other states is possible (and perhaps more probable), is the achievement of "order." It costs us considerable work to maintain order, and the larger the number of other possibilities, the harder we have to work. The universe does not discriminate except in preference for equilibrium, which is simply a consequence of cumulative probabilities. Our discrimination in favor of order is the triumph of conscious life, the capacity to organize, the freedom to choose, the ability to perceive and communicate.*

* * *

OUR EXPERIENCE OF a unidirectional time, the following of cause with effect, proceeds from the irreversibility of processes "far from equilibrium," as explored by Prigogine and Stengers. [3] Without purpose, the universe indeed "winds down" toward equilibrium everywhere, all structure decaying. The presence of intelligent life introduces purpose into otherwise random processes, however, sustaining far-from-equilibrium organization through the conscious application of directed energy. The genius of life is to sustain the improbable. The conscious recognition of time is a necessary consequence.

Stephen Hawking, in his elegant essay *A Brief History of Time*, names three different "arrows of time": the thermodynamic, psychological, and cosmological directions. [4] The thermodynamic arrow, nature's downward slope, is the universal experience of spontaneous disordering: time moves in the direction of decay, of work dissipating into heat, of increased equilibrium and sameness. The famous second law of thermodynamics proclaims that most processes are irreversible, that it takes more energy to restore or maintain order

---

* Order is not, of course, inherently benign. Intelligence is expressed in bad behavior as well as good. Our ability to organize has produced exploitation, addictions, weaponry, and the myriad illnesses of crowding and greed. The achievement and sustaining of order is a necessary but not sufficient condition of life, which can prosper only through the positive expression of its qualitative aspects. (See the letter from the South Bronx.)

than to lose it. The measure of disorder is called entropy, and it can only increase. Hawking points out that our psychological experience moves in the same direction of time (effect following cause), and that our universe is expanding as the same time passes. Why should all three arrows point the same way? Thinking humans could survive only in an expanding universe, Hawking says, and their thought processes are subject to the same thermodynamic rules as the world around them.

I would add a fourth arrow of time to Hawking's three: the direction of accumulating information. Our time advances with the increase of information, and information is not "conserved" in the sense of physics; that is, information can be created, much as entropy is seen only to increase in each (irreversible) transaction. (In fact, Shannon's measure of information, which he named entropy, is mathematically equivalent to thermodynamic entropy. [5] Both grow inexorably in our universe, and thinking beings add to them at an increasing pace.) New information is found in discovery and shared in communication. The store of memory thus created only grows, and the direction of its growth gives us our sense of time. (Of course one can forget, but if one should forget everything, one would indeed lose track of time.) [6]

The physicist and writer Heinz Pagels makes a similar argument for the memory arrow of time:

> The increase in entropy implies that time has an arrow—there is temporal irreversibility and physical processes exist which can store information; memory is possible. [7]

Only the disordering of irreversible processes and the equivalent increase of information, in all our normal experience of physical occurrences, has an intrinsic time-direction. All other occurrences are entirely reversible, and make as much sense backward as forward.

Cause would be indistinguishable from effect, regression from progression, without that ever-present tendency toward disorder, toward
equilibrium. The universe's proclivity for equilibrium is simply the
descent toward the most probable conditions and arrangements of
all things. The nobility of life—even the possibility of conscious life
at all—derive from our capacity to maintain the Prigogine-Stengers
"far-from-equilibrium" condition. That condition depends on our
capacity to build and maintain order, to extract it or perceive it in
the midst of otherwise disordering processes.

* * *

LIFE IS THE GREATEST of improbabilities. The evidences of life
with which we are familiar appear to result from chance and natural
selection; does this make our freedom to choose an illusion? Many
things in the universe organize spontaneously, notably upon cooling,
and all things emit some information, notably upon heating. Few
things get to choose.

A test for free will, Stephen Hawking says, would be non-pre-
dictability. [8] If one can predict every move, the presence of freedom
is unlikely. And although one cannot predict human behavior, the
reason is not a free will outside the laws of physics, but the stupen-
dous complexity of possible events. So all things are determined,
Hawking says, but it's all so complex that it doesn't matter.

Although it seems unlikely that all our choices and behavior
could be predicted even if all the data could be gathered and all the
equations solved, one does not lightly dismiss a Hawking conclusion.
Indeed we need to understand freedom within the laws of physics—
i.e., in a rational context.

Absolute determinism, of course, is rendered obsolete by the
quantum theory. The Newtonian concept that God wound up the
world like a clock and everything that follows is predictable (although
complex) is discredited by Heisenberg's uncertainty principle and the

unpredictability of random events in the micro-world. Precise information about the states and future of elemental particles is simply not available to us, and this is not a failure of measurement but a fundamental fact of nature.

Although the indeterminacy of quantum events means that all is not determined in advance, there is still the question of the degree to which intelligent life can choose, can affect future probabilities. Part of our problem has been in seeking precise boundaries again, trying to defend a human distinction from animals, defining a God-given free will that one has or has not, with no shades of gray. Freedom, however, as a capacity to choose in the absence of (or in spite of) constraint, develops gradually in human evolution, varies widely in all of us from time to time, and generally grows in collaboration.

Freedom is "becoming" as we advance. The simplest one-celled bacterium can follow a chemical trail, choosing a path toward nutrients. The migrating bird can combine topographical, celestial, and magnetic hints to choose a way home, and the spawning pacific salmon chooses sex over food to perpetuate its species at the expense of its own death. Primates and early humans developed gradually more complex combinations of signal-response behavior, until today we can make very abstract choices whose stimuli are not so identifiable.

As humans have evolved and matured, these stimuli and influences on behavior have become more numerous and complex—from chemical and hormonal balances to instincts to primitive communication, on eventually to the ability to organize, to recognize and build order, to relate, to cooperate, to commit. All these drivers and capacities go into our decisions, our actions and choices. The higher motivators of relation and commitment often, but not always, dominate the lower motivators of appetites and instincts. The hardware of our genes and then the software of accumulated experience overcomes the instinctive response.

At some point in this evolution—or in this cycle, for we regress and vary in its expression—the motivators and responses take on what we consider to be free will. No laws of physics have been broken, but our identities have become more expressed and more intertwined. This becoming continues today. Evolution is not arrested, and in fact may be considered to accelerate as we develop greater means of communicating and working together. The expression of intelligence, whether the familiar human kind or other kinds likely in our universe and others, naturally progresses with the sharing of information.

The dismissal of free will is based in reductionism. Elemental particles have no choice, and we are aggregations of them. But are we nothing but such aggregations? Here we need to look more closely at the distinction between the micro world of elemental particles and the macro world of large aggregations such as ourselves and the objects we handle. Again a bright line of distinction cannot be drawn and defended against all logic. A qualitative difference has been pointed out, however, by Heinz Pagels: it is the capacity of memory.

Memory is possible because of the nature of information. When an observation is made—by ourselves or by another life form or by any system with memory—information is obtained. The storing of such information is possible only if memory exists, and memory exists only if the arrow of time is sustained: no past means no memory.

Elemental particles have no memory; intelligent systems do. The irreversibility of time and observation allows us to accumulate information, to organize, to learn and build, to communicate, to work together toward common purposes, to share. This, combined with our growing freedom, is what we call life.

* * *

WE ASSEMBLE MACHINES in our lab, and once assembled, each machine is nothing but a particular arrangement of its parts. It can be faithfully reproduced from our drawings and specifications, the digitized storage of which contains all the information necessary. I cannot, however, assemble a different machine from the same stored bits—nor can I compose a 10th symphony from a CD of Beethoven's 9th, nor a chimera from a cloned sheep. What makes this so hard? I cannot find the design process among my bits, or Beethoven's plan among his notes. The final product's memory includes only the end result, not the creative process. The information used in that process is partly lost in waste products and partly remembered elsewhere, but it is not available from taking the product apart and analyzing it.

Michio Kaku points out the need for both the "holistic" and the reductionist views in understanding physical reality, showing that neither is complete and that one may reach the same conclusions by either route. [9]

To predict all behavior would require holding all information, past as well as evolving. This is not just extremely complex but infinitely so, because information keeps growing and we are contained within the puzzle we would solve. It would require all intelligence working perfectly and forever together, in complete dedication to the task and to truth. Only thus could the puzzle be solved, by Truth itself and the Whole Each Other.

Notes

[1] Annie Dillard, *Pilgrim at Tinker Creek* (New York: Harper & Row, 1974), 68.

[2] Hebrews 11:1.

[3] Ilya Prigogine and Isabellet Stengers, *Order Out of Chaos* (New York: Bantam Books, 1984.)

[4] Stephen W. Hawking, *A Brief History of Time* (New York: Bantam Books, 1988.)

[5] See the letter from the Radio Communications Laboratory, and its references to Shannon's information theory.

[6] This basis of time in the flow of information is explored further in the letters from Tilden Pond and The Board Room.

[7] Heinz R. Pagels, *The Cosmic Code: Quantum Physics as the Language of Nature* (New York: Penguin Books, 1994); 123, 140–141.

[8] Stephen W. Hawking, "Is Everything Determined?" in *Black Holes and Baby Universes* (New York: Bantam Books, 1993.)

[9] Michio Kaku, *Hyperspace: A Scientific Odyssey Through Parallel Universes, Time Warps, and the 10th Dimension* (New York: Anchor Books [Bantam Doubleday Dell], 1995), 316–328.

Manhattan Island

I N THIS LETTER I escape from a traffic jam on FDR Drive, along the East River heading north from the Brooklyn Bridge, which my taxi driver took in a failed experiment. This does not constitute a better way around, and we are both in foul moods. It is raining and cold along the hard asphalt, darkening the graffiti-covered concrete, closing in the already short horizons, adding noise on noise, turning the metal oxides in the heavy air to acid. In this small prison I am late and getting later, fretting about the lawyers waiting for me downtown (impatiently I am sure), trapped in a palpable vice of frustration.

Now we are entirely stopped, and with nothing to look forward to I look to the side. There is my image in the window: there the scowl, below it the narrowed eyes, the tight lips, the clenched jaw.

And there in the center of my image, improbably, appears a small flower. I focus on it: the flower grows in a tiny crack in the pavement, near a crumbling curb, a lone touch of color in a gray world. Defiantly it affirms the potential of life, the possibility of beauty. My jaw releases and I remember commitments older than today's meeting.

* * *

W E SEE SO LITTLE. Partly this is a lack of training, partly of aptitude, mostly a failure to look. For this reason, among others, we need artists. Driving past St. Patrick's, I look at the cold stone and see the color gray. I would give the same answer if asked to describe Rouen Cathedral, but Monet knew better. His famous renditions of its facade in 1894 used oils of blue, gold, and rose to reveal the changing character of his subject in the changing light. His expressions evoke from our experience a conscious and subconscious recognition, an integration of pieces of memory, a host of connections and realizations we had not noticed or assembled before.

Monet could paint because he could see. My vision, so much poorer, misses much until aided by the artist—and even then it struggles to discern what has been realized by the critic, the student, the more astute observer. Probably no perusal captures all of Monet's meaning, and probably even he could not tell us the whole idea, and so he continued to paint.

I have the same problem as I write. My ability to describe what my eyes have seen is deficient, a deficiency not of drawing but of seeing. I do better at what I have heard or contemplated. Annie Dillard, by contrast, is a "visual" writer. She can put into compelling prose what her eyes have taken in—because she can see. As a girl, Dillard spent hours drawing and studying the visual arts and the natural world. She built a thousand resonances in her mind, tuned sensitivities that ring when she stops to absorb a field or a brook or a shadow—things that seem ordinary to me until she describes how special they are.

* * *

IN ANOTHER APRIL I have seen the same color that startled me on FDR Drive, multiplied by millions, along another well-traveled road. Dallas and Houston are separated—no, connected—by 250 miles of wild flowers. Most prominent are the bluebonnets in the early weeks, interspersed with orange paintbrush. Driving south, they yield to white, yellow, and maroon species I cannot identify. The colors slope up from the highway and roll across the prairie, sprinkling the young grass beneath the oaks and cottonwoods, tumbling over one another, competing for the space left green, heedless of the time that is so short for them and us.

Driving through them under a blue sky, one tries to absorb and retain the beauty hurled at us by each mile, but it is too much. We are quickly overwhelmed by the impact, like the surfeit of Monet at the Musée Marmottan or the cascading *Dies Irae* in Verdi's Manzoni Requiem; we try to slow down, to open wider, to remember, but the moment flees.

Perhaps we are simply moving too fast; if we could slow our pace, find a quieter place where we could reflect and contemplate, perhaps we could take in beauty and understand it. There is a Vermont night I know, just after a heavy snow, under a half-moon that had thrown a halo of ice around her in the sky, a deep cold night in which nothing stirs: it is absolutely hushed, suspended, silent as death. There is not the slightest breeze. I strain to hear, and only the inside of my own head answers. My mouth opens, my breathing stops, in awe. I despair of description; we do not have the adjectives and I cannot find the metaphors. I remember Annie Dillard's awe at the edge of a farm field in a different season:

> ...suddenly, I saw the silence heaped on the fields like trays. That day the green hayfields supported silence evenly sown, the fields bent just so under the even pressure of silence, bearing it, palming it aloft: cleared fields, part of a land, a planet, that did not buckle beneath the heel of silence, nor split up scattered to bits, but instead lay secret, disguised as time and matter as though that were nothing, ordinary— disguised as fields like those which bear the silence only because they are spread, and the silence spreads over them, great in size. [1]

The young Robert Frost, confronted by an apple orchard in spring, tries:

> Like nothing else by day, like ghosts by night...

then bursts with the impossibility of containing it:

> For this is love, and nothing else is love,
> The which it is reserved for God above
> To sanctify to what far ends He will,
> But which it only needs that we fulfill. [2]

It does feel like love. As I stand under the winter moon, that is the only word that comes.

* * *

So this is our experience with slower beauties. Even when time accommodates study and reflection, letting us turn it around and contemplate it, beauty will not reveal its identity. It simply stretches out to match our time. The lone brave flower on the East Side is as elusive as the flood of color on the Texas prairie.

On either road I can stop and dissect a single blossom, probing for the source of beauty: surely it is in there somewhere. But it eludes my probe, somehow slipping through my fingers, as it does when I examine Monet's brush strokes or Verdi's notes, or the spectra of stellar radiation from a Vermont sky.

I do not draw from this elusiveness, however, the conclusion that beauty cannot be understood in rational terms. Of course the reliance on reductionism is inadequate, the knife of dissection obliterates connection, information is lost in disaggregation; and of course we cannot define without limiting. Yet I believe a rational understanding of beauty is necessary if we are to share it fully. The same is true of love, and of life. All depend on communicating, which succeeds in proportion to information shared. Life is not sustained by mystery, by inadequate or inaccurate information; nor is love requited, nor beauty discerned. Any beauty that is diminished by improved vision was, I propose, illusory.

This does not mean that beauty is captured by discernment; it goes deeper than that. No description or measurement fully defines the reality of any phenomenon under observation, even if it is as "quantitative" as a particle's exact location and energy; yet we have made great strides toward verifiable theories of most physical realities. Admitting our inability to capture, to define, to measure, or to comprehend anything in absolute and complete terms, we yet seek to understand it better. Such seeking should be as earnest in pursuit of beauty as in pursuit of any other manifestation of truth. Even as we fail to pin them down, we should be capable of increasing our understanding of things beautiful; and since beauty lives in our perceptions, such an increase should enhance our appreciation.

Why bother to understand beauty? To share it, I think; to benefit from its evocative powers, stretching us beyond declarative prose; to communicate more efficiently, and thus "become" together.

* * *

To TEST THESE ideas, your mom and I have obtained a rather special pass to an unusual kind of art museum, rarely visited. There we hope to find substance in the resonance we feel when struck with ancient art, the faith that physicists so often express in elegance, and the connection of beauty to life.

Notes

[1] Annie Dillard, "A Field of Silence," in *Teaching a Stone to Talk* (New York: Harper & Row, 1982.)

[2] Robert Frost, "A Prayer in Spring," in *Collected Poems of Robert Frost* (New York: Halcyon House, 1939.)

Hard Scrabble Wash

Few ANGLOS AND not many Native Americans can lead you to Hard Scrabble Wash. Crossing the uniform plain of northeast Arizona it is invisible from more than a hundred meters in any direction, startling and alien when encountered. The wash is a violent parting in the desert floor, an ancient fracture of hard sandstone, now a hundred feet deep and once deeper. At its beginning the fissure is only a few feet across, but it widens rapidly into a canyon, twisting and spreading for a mile until it empties again into the desert plain. For most of its length the rock walls are nearly vertical, defying escape.

Along this tributary drainage of the Zuni River, the earliest pit houses were inhabited around the beginning of the Christian era, marking the gradual ascendance of agriculture over the hunting and gathering economy that had prevailed in the millennia since humans first crossed the ice-age bridge from Siberia.

We slide and worry down one of the few rock faces that accommodate human descent, holding on and watching for the rattlesnakes whose home we are invading. Our objective is the deepest recesses of the narrow end, which hide a treasure of petroglyphs, many of which have never seen the sun. One must squeeze between rock walls and wade through a dark pool to enter their cave, and bring a light to see them. But the reward is an unspoiled, varied, extensive display of ancient art, impressive beyond the more popular sites.

This is the only written record of a people who largely disappeared before Europeans began their exploitation of the continent. The Navajo call them Anasazi—those who moved on—and their descendants are probably the surviving pueblo tribes, notably Zuni and Hopi. All the painted forms that have been found elsewhere are here in the cave and down along the steep walls of the wash, many with clear symbolism and others still mysterious. At the beginning of the wash they are all men and animals and symbolic figures, but not yet corn: that is the subject of the later paintings farther down the canyon. They are religious and artistic as well as historical, and strike a resonance with our own strivings.

The resonance is more than pattern recognition. We hear a familiar music in this early art, this hope, this prayer, a haunting music that is still sung. We can recognize the melody in church and in prison, in the laboratory and the hospital, on the rude streets and on the mountains, in places of power and despair, of brilliance and darkness. It is sung in chorus and duets and solos. The notes can be spread on canvas and danced and run, and they can be precipitated as crystals in poems.

Indeed, poems. We and the Anasazi can build roads and raise temples and cultivate the ground, but it is our dreaming and wonder and search for longer connections that make us one. Our children will synthesize everything in sight but still struggle with poetry—with meaning and order and relation, with communication.

* * *

As we puzzle over the meaning of the symbolic forms, your mom reminds me of the more complex symbols left by the Anasazis' cousins far to the south. Mayan and Aztec artists, historians, and priests evolved a symbolic language of increasing abstraction, as have other peoples. Over time, abstract symbols for sun or snake or crops or weapons enter the shared language, carrying an ever-increasing weight of meaning.

A symbol begins as an obvious drawing, an analog to a familiar object, but it evolves. After some generations it may become unrecognizable as the original object, but to the initiated it will convey immediately not only the object's identity but its cultural, religious, historical, or political significance. It conveys more information because of the stories and practices and shared experiences of the people who write and read it. It becomes a code, not just for the object but for what the object may represent in the observer's culture. Eventually whole histories, rituals, personages, and concepts are suggested by a single stroke. Like that story about the inventory of jokes endlessly repeated among a prison's population, all we have to do is shout out a code number and everyone laughs.[*]

This shared context is important because the efficiency of communication grows along with the sophistication of our symbols. Whether in language or in art, each symbol strikes within us a resonance, tuned by both personal and shared experience. So a poem or a dance does not attempt to spell out its message in linear form, but exploits such resonances to communicate with greater reach and longer effect. Poetry and other art forms reach varied audiences in ways that are similar but not the same for all. As a pure note at the right frequency will set a particular crystal glass ringing, a line of poetry will find its welcome in different readers at different times. I think this is true of all the arts.

The arts express the inexpressible by striking such resonances. Prose is intentionally declarative, definitive; I consider art to be that form of expression which neither declares nor defines, but evokes. When the resonance is struck, by word or art or mathematics, much information is conveyed efficiently.

---

[*] The punch line, for those who haven't heard it, comes from an inmate who laughs unusually hard at "23J." Asked why this selection from the repertoire strikes him so funny, he replies that he had not heard it before.

Beauty thus gives us a kind of shortcut to truth. It is an efficient form of communication. Anyone who has done serious science or art knows the psychological experience of beauty inspiring discovery or creation. This is not mystical but rational, having its base in communication. Beautiful things are evocative of hidden connections, of unformed thoughts, of new ways to see. They call forth insights, perceptions, or correlations that were latent but not before articulated.

* * *

Is THIS TOO cold, too calculating, too mechanical? Dare we explain beauty anyway? Would that not miss its impact, detract from its power? Beauty, it is said, appeals to our spirit, not our logic. If I turn upon it the cold analytic eye of reason, I see only the grains of its structure, not the genius of its wholeness. How shall we explain a sunset, define a child's smile, analyze the grace of a dance or an athlete, describe the cry of a loon on a summer night, declare what makes the woods beautiful or the stars to sing?

* * *

ANNIE DILLARD, MORE schooled than I in both letters and graphics, writes of the song of a mockingbird on her chimney:

> Why is it beautiful? I hesitate to use the word so baldly, but the question is there. The question is there since I take it as given, as I have said, that beauty is something objectively performed—the tree that falls in the forest—having being externally, stumbled across or missed, as real and present as both sides of the moon. This modified lizard's song welling out of the fireplace has a wild, utterly foreign music; it becomes more and more beautiful as it becomes more and more familiar.... Beauty itself is the language

to which we have no key; it is the mute cipher, the cryptogram, the uncracked, unbroken code. [1]

So, Dillard says beautifully, beauty exists whether we perceive it or not; beauty increases with familiarity; and beauty cannot be understood rationally. In a paragraph she sweeps away three of my principal hypotheses about beauty: that it exists in the perception of intelligent beings; that its communicative value diminishes with repetition; and that it can be understood in rational terms.

Is she right and I wrong? Shall we give up on the rationality of art and consider it, as many do, beyond objective understanding? Perhaps we should leave mysteries alone, reserve some space for mysticism, like a forest preserve or conservation land, protected forever from the rude inquiry of rational thought. Perhaps we should hold that some things cannot be explained, that beauty and love and other things spiritual are given, impenetrable, inherently out of reach.

Or shall we hold that to seek greater understanding of anything—or anyone—is an act of respect; that one can appreciate more the depth and meaning of those things that one takes the time to study? Wonder and awe, we might observe, only increase in the search for truth. The respect and love of beauty may be enhanced by seeking to understand it. We do not expect to grasp the whole reality of anything, for nature does not permit the holding of complete information, but the beginnings are accessible and the effort is faithful to truth.

I remember emerging from a late class on a winter evening, startled by a full moon rising over campus. As an engineering student, I could calculate the time those photons took carrying their image to my eyes, I could describe the orbit of our satellite and I knew what made it look full. Yet none of my studies came to mind, and my reaction was as emotional and romantic as any young lover who had never attended a math class. My response was impulsive, almost instinctive, not intellectual. So why have humans made and sought beauty long before they could define its media or analyze its subjects?

* * *

WE CONSIDER THINGS beautiful in which:

- We perceive a hidden connection, an underlying order, a unity in variety
- We see the possibility of freedom
- We find a new representation or view, we see or hear or feel more deeply

Does this sound familiar? We proposed exactly these evidences for intelligent life—the discernment of order, the freedom to choose it, the ability and proclivity to communicate. Our instinct to pursue beauty may therefore be rooted in its transparency to life. Beauty and love seem to me our great windows on life and its possibilities.

* * *

SUBRAHMANYAN CHANDRASEKHAR, an astrophysicist, is quoted as saying:

> This shuddering before the beautiful, this incredible fact that a discovery motivated by a search after the beautiful in mathematics should find its exact replica in Nature, persuades me to say that beauty is that to which the human mind responds at its deepest and most profound level. [2]

This exclamation, and its quotation by other scientists, expresses a kind of faith that has time and again been rewarded in scientific exploration. The laws of nature are predictably expressed in mathematical statements that are beautiful (elegant, to use an adjective that captures efficiency of expression as well as symmetry of form.) Physicists use elegance as a reliable guide in seeking new understanding. [3] The moment of discovery, when the beautiful in mathematics leads to successful theory, when the theory successfully predicts physical reality, evokes the same exultation and awe as discovery in art. It seems to me not at all a sacrilege to liken this to a religious experience.

Beauty is perceived when we are able to extract order from chaos and to discern truth in a new way. It is an associative capacity, linking objective observation to subjective concepts. To use the Navajo saying, we can walk in beauty. Our ability to communicate perceptions is very limited, and beauty gives us a way to share experience.

* * *

So we may dare to understand what beauty is. We dare not, in fact, set it apart from our rational pursuits, or it would be like another religion of antiquity, which we are afraid to question because it is so fragile.

Truth—the intrinsic reality that prevails in every event[*] and integrates all information—is likely infinite, and we shall not capture it except in fragments. I consider beauty, however, to be a capacity of intelligence, as are life and love, to which it is closely related. The capacity to recognize or to bring order out of chaos, the freedom to do so, the ability to make commitments and to communicate—these capacities are what we hope to hear on Harvard's hill. That we are given such capacities as a people seems to me a sacred trust, and beauty a way to celebrate and share it.

---

[*] I have used the word "event" in these letters in its relativistic sense (generally speaking) of a point in time and space. "Every event" would include all possible points, even those beyond the observer's "event horizon" (those she cannot see or affect.)

I propose that beauty exists in its perception; which is not at all to say that it is tenuous or unreal, but the opposite, that it is related to life. The universe's gift of life and beauty is the capacity to extract order, in creation or in perception. We can find it here in Hard Scrabble Wash, in the Louvre, in a woods or a symphony— and the wonder is, we could find it in a page of tracings done by a random number generator. I can program this computer to do just that, post the result in a museum, and many will admire it. That fact does not suggest that beauty has no meaning but that it has: it celebrates our capacity to find order in what is by definition chaos; to extract from purpose-less displays some form, some balance or structure, relationship or metaphor. For such a capacity we should be far more grateful than for a mere acknowledgment of intrinsic, given, pre-defined art forms.

If there were intrinsic beauty, it would be defined and pre-ordained, infringing on our freedom. We may feel some criteria are important—balance, symmetry, harmony, internal consistency, for example—but every rule reduces the artist's freedom and the audience's openness. Some context is essential to communication, but that is not the wonder of beauty but its limit in our hands.

Contemporary music, poetry, and visual arts tend to explore the potential of an internal reference in place of context. That is, novel uses of tone combinations, words, or images may have unconventional meanings, known initially only to the artist. In such cases, the listener (reader, observer) must find the internal reference before she can fully appreciate the work. This sacrifices some of the immediate impact that more conventionally-based compositions have, in the hope of expressing wider or deeper meaning. The internal reference becomes the context. The contemporary artist exercises greater freedom, includes more surprise in the composition, thereby offering more information, a richer work than he could have built within traditional bounds. The offer, of course, is not so easily accepted.

Often we as the audience are impatient with this process, preferring the greater comfort of familiar structures. Indeed, given the constraints on our time and apertures, we may be unable ever to capture the messages intended by our contemporary artists. Some of them will be appreciated only by future generations who will have the benefit of longer study and accumulated familiarity—building the shared context. Some will never be fully understood; and some will be found to have said very little in spite of mighty effort.

Of course some works of Beethoven and Blake and Monet were greeted with the same impatience, even derision, in their time, and so it must always continue if art is to live. Kurt Vonnegut Jr., who I suppose had to struggle a bit with his own patience in appreciating the art of his daughter and her contemporaries, often involves visual artists as sympathetic characters (a rare distinction) in his books. Here he gives his minimalist artist Rabo Karabekian the last word in a cocktail lounge in Midland City. [4] Karabekian has sold the city a painting titled "The Temptation of Saint Anthony," a 20 x 16-foot composition with a single vertical stripe against a uniform background:

> "I would love for your children to find pleasantly and playfully what it took me many angry years to find," says Karabekian. "[My painting] shows everything about life which truly matters, with nothing left out. It is a picture of the awareness of every animal...the 'I' to which all messages are sent. It is all that is alive in any of us...unwavering and pure...A picture of Saint Anthony alone is one vertical, unwavering band of light. If a cockroach were near him, or a cocktail waitress, the picture would show two such bands of light. Our awareness is all that is alive and maybe sacred in any of us."

It is hazardous to speculate where Vonnegut is speaking for himself and where for effect, whether he is leading or pushing us, which of his serious pronouncements are fact and which fancy, as he glides with his unique facility in and out of his books. Yet he communicates with great elegance, and his speaking for the visual artist or the scientist, both of whom he knows up close, is likely to be insightful.

Vonnegut's Karabekian and my random number generator—and every artist who claims to create only for herself—communicate in spite of themselves. There, in fact, we may have a criterion for judging art: its ability to communicate. A composition that conveys no surprise, no new information, seems to me less a work of art than one that gives us insights or revelations or discernments or new ways of perceiving or considering. The artist sees for us and shows us what she sees, after which we see more.

The artist may not begin a work in order to communicate outward, but works because he is assaulted by ideas or colors or sounds, striving to express his own relationships with his own world. Monet says he was overwhelmed by variations only he could see in light and color, the swiftness of their change, the intensity of their flow. The artist is often close to madness, always in despair that her experience cannot be understood, by others or herself. Yet this is communication too: in the agony of discernment, the labor of birth brings forth to us all a new view of life. The artist bursts and delivers, and we are joined.

* * *

THIS IS THE capacity to see. This is the link of beauty to life, to seek together a broadening of the context within which our freedom and understanding can grow. To preserve such freedom is to be committed to life as well as to truth—and to each other.

Notes

[1] Annie Dillard, "Spring," in *Pilgrim at Tinker Creek* (New York: Harper & Row, 1974), 106–107.

[2] Quoted in Heinz Pagels's *Perfect Symmetry: The Search for the Beginning of Time* (Simon and Schuster, 1985), page 54, and re-quoted by Michio Kaku's *Hyperspace*, page 226.

[3] See, for example, "Is beauty necessary?" in Kaku's *Hyperspace*.

[4] Kurt Vonnegut, *Breakfast of Champions* (New York: Dell Publishing Company, 1973.)

The Massachusetts Institute of Technology

O N THIS CAMPUS where the buildings are named for numbers instead of people, I look out at my audience of nascent
engineers and wonder: who among you would agree that
you have been studying the means of art? the philosophy of civilization? the capacities of life? No hands would go up, I fear, or perhaps
someone would say: is this a trick question?

I'm on a panel this morning at Kresge Auditorium, sharing
the stage with more prominent speakers, discussing alternate career
paths for the engineering graduate. My invitation was probably
based on my recent resignation from a defense contractor to start a
company trying to apply technology to social needs, and on being
president of the Boston Industrial Mission. The remarks made here
this morning will be captured and edited by Jonathan Allen and
published by the MIT Press. [1]

Hopefully my remarks are inspiring, but I am not an exciting
speaker, and earn polite applause. The large audience is waiting for
the more impassioned oratory of Nobel Laureate George Wald, and
they will not be disappointed. Also speaking are laureates Hans
Bethe and Salvador Luria, and 22 other biologists, physicists, business and academic leaders, assembled here by a common concern
for the direction of what President Eisenhower named the Military-
Industrial Complex.

* * *

IF I WERE as prominent as my follow speakers, or perhaps just more mature, I would have asked this of my young listeners: As scientists and engineers, can we hold anything "sacred"? Is there something, rationally speaking, to which we can devote our careers with confidence in the worth and meaning of what we do? Can we derive from first principles something to which our profession should be faithful?

You might answer "to truth, wherever it leads," and I would say yes, that is special in our profession; but truth prevails in every event, whether we are faithful to it or not. What else needs our commitment to do our work? Logic directs us to consider freedom as an object of scientific faith; i.e., we must minimize the constraints on our exploration, on the hypotheses we adopt, and on our sharing of ideas. History is crowded with examples of such constraints, impeding the progress of science.

Given the prevailing of truth and the maintaining of freedom, I think you would also propose some special capacities as necessary objects of our commitment—capacities that have been sorely taxed here at MIT but which have carried you through. One is the ability to exchange information, to communicate. Although some discoveries and inventions appear to have sprung spontaneously from individuals working alone, the application of their work, and the accumulation of information leading up to it, required communication with others. Moreover, the working of our brains themselves, and all their support systems, require exchanging information continuously at the cellular level. If communication were no longer possible, learning would cease, as would living.

Finally I think you would propose the capacity to organize that information, to recognize and bring about order. This is the capacity to recognize patterns, to make energy into work, to reduce entropy locally, to arrange and structure and build, to learn and teach and heal, to organize.

These four—the pursuit of truth, the maintenance of freedom, the exchange of information, and the recognition/bringing of order—I propose as the rational core of our devotion. They are also the central genius of what we in the sciences would recognize as intelligent life. To me, this is a beautiful as well as a rational realization. These capacities support our ability to build, to organize, to correlate and discover, to grow together, to see beauty and make it, to share, to love, to become. Is this not the goal of our profession? Our devotion to the continued development of this ability—and to its highest use—is the most sacred commitment I can imagine.

* * *

THERE YOU HAVE it. You have derived a rational faith from first principles and self-interest as engineers and scientists. The objects of our trust and commitment are truth, freedom, and the capacity to communicate and organize. Not coincidentally, when we search for signs of intelligent life elsewhere in our universe, it is the combination of that freedom and those capacities that would constitute detectable evidence. So our faith is in truth and in life—not a surprising conclusion. Its implications, however, are great. Just as we are committed to truth wherever it leads, we have derived a commitment to intelligent life, *whatever form it may take.*

Whatever advances freedom, communication, and organization advances intelligent life. This raises the possibility that you and I may not be the most highly evolved and special bearers of this extremely low-entropy capacity—or to put it more hopefully, that we have yet a long way to go. Our particular evolution, after all, is a pretty recent event in cosmological time. On our planet, evolution has taken the familiar form that we call biological—at least until perhaps the afternoon of the nineteenth century. You live early in an epoch when this is no longer true, when our growing capacity to organize and communicate—and our freedom from ancient constraints—will be based on your inventions.

I know this is unsettling to many, but consider this: What can one find in the biological evolution of homo sapiens from the mud of earth that confers upon our life-form either distinction or holiness? Unless one subscribes to the Genesis account, the gradual development and specialization of DNA and its expression seems not, per se, superior to the building of intelligent robots out of Tinkertoys. We like to consider the human accident (as Gould would call it) a very special and somehow noble event. Thus we cringe at the suggestion of non-biological intelligence and the replication of human organs through genetic engineering—yet a well-guided, purposeful "creation" of this type seems to me not necessarily less holy or teleological than the stumbling, accidental evolution of the human species.

This is not to be cynical but in fact to be hopeful and respectful. We are capable of beneficent purpose, of commitment to truth and freedom and each other, in this wildly improbable capacity that we have inherited to organize and build. What we choose to construct, in such a commitment, can be more sacred than any accidental accretion of hydrocarbon molecules, because it is imbued with coherent purpose.* And the admission of more than one expression, form or development of intelligence is not inconsistent with the possibility of an intrinsic, primordial concept of Intelligence itself.

We must admit that the presence of "life" (unless narrowly defined in an anthropomorphic tradition) is neither distinct nor invariant. Intelligent life exists in shades and degrees, and varies in all of us. As we increasingly augment our ability to organize and communicate, and our freedom to do so, we increase our expression

---

* Of course we may fail. One can be faithful only to one's self. But the rational choice of mutual benefit has so far prevailed in the long run, in spite of vast sweeps of dishonesty and cruelty.

of such life, even though the augmentation may be non-biological. Capabilities not derived from sexual intercourse are not innately artificial. (I hesitate to make such a claim on a coeducational campus, but I have studied this longer than you.) In any event, you will surely extend the reach of our species and push out our boundaries, through the increasing confluence of our biological and electro-mechanical streams of development.

This faith in truth and each other stands without dogma or mysticism. It is shared with the artist, the philosopher, and the enlightened leader, all of whom require the same principles to move forward. Your self-interest is the world's interest, and its interest is yours. In such a commitment you can find meaning in your life; on such a commitment you can build your identity.

* * *

HERE AT MIT some of the longest strides in the communication sciences were made. As we ponder the impact of our work, we might speculate on where communication could take us as a people. First, it seems unlikely that we have exhausted all its means. The media developed by carbon-based life forms on planet earth are chemical (smell, taste), tactile (acoustic, vibrational, touch), and electromagnetic (sight, telephone, radio.) These rely essentially on our modulation and detection of photons, electrons, and larger particle aggregations, which appear to be universally available. This has supported our conceit that communication would be effectuated everywhere via the same media, which is less likely.

If consciousness exists in the universe beyond the functioning of discrete brains like yours, it should support information transfer and sharing generally. Speculating on the use of other media is of course hazardous. Life, however, is a tenacious and aggressive thing, and at some point seems likely to break free in any context, to develop the capacity to organize, to choose and to communicate.

These capacities underlie the progress of civilizations, the local growth of order and its proliferation. Any means of communication would likely be seized to advance such progress, and many means may exist.

We have found ways of transfer and sharing that suit our particular estate. Other life forms, if there be any (which seems likely), might well develop other means—other media as well as other symbols. Everything above absolute zero temperature transmits, and everything with any aperture that is below infinite temperature receives; but only those blessed with freedom and the capacity to discern order live. Wynton Marsalis advises, in life as in music, to go "to the edge of the possible, without losing balance." He is describing what distinguishes "swing" in music, balancing—as we must in all forms of communication—uncertainty with context. The percussion sets the context with a "shuffle" rhythm of 4/4 vs. 3/4, while the horns challenge the beat ("chasing it") and experiment, adding the uncertainty that Shannon tells us quantifies information transfer. [2]

When you leave MIT, will you aim for the edge of the possible? To the extent that we share information we share experience, and by extension share our selves. Identity begins—but does not necessarily end—in the things by which we are recognized. Superficially they are nominative (my name, my credentials, my profession); historical (my accomplishments, failures, reputation); and sensory (my appearance and "feel" in the media we share.) At a deeper level they are relational and predictive—expressions of commitment, transparency, relation. It is at this level that we can share our selves, that we can stretch the edges of our possibilities without losing balance.

You have studied the history of science and civilization, and recognize that the interests of "self" have best been served by a commitment to truth and to others. To these means of maximizing the depth and meaning of our profession we might add a third: to grow or advance together, to stretch our expression of the life-capacities, to "become" as a people. This requires efficient, accurate communication. Indeed one cannot "own" intelligent life, but can only share it. We lose life when we try to own it, to hang on to an isolated identity. We gain life by sharing it; i.e., by building and communicating together, by enhancing each other's freedom. A wider sense of "self" accommodates a broader self-interest. Things live only by the association of components, of participants.

* * *

FROM THIS STAGE in the late winter of 1953, a series of three lectures was delivered to a similar scientific audience by Carnegie professor Jacob Bronowski, lectures that eloquently explored The Creative Mind, The Habit of Truth, and The Sense of Human Dignity. [3] Bronowski discussed not only the deepest meaning of science, but argued cogently that the best principles of science are also the underlying principles of the arts and civilization itself.

It is apparent that our capacity for order grows as we join in common enterprises, and thus we may be said to be enhancing life in proportion to our connections. The presence of art, humor, communication, and organization characterizes our civilization, and probably other civilizations that take forms presently unknown to us.

Life is, as are all realities, a matter of probability. The propagation of periodic crystalline structures in many media is quite orderly, but lacks freedom and purpose. The propagation of "biological" molecules appears more purposeful; yet biological life, as we know it on this planet, springs originally from apparently random events. As we evolve and find common purpose, however, our ability to alter the probabilities of future events grows. As we are free of bonds we are free to join. The commitment to life is a commitment both to enhance and to share.

Intelligence appears to arise from natural selection and accident in evolution, which are highly probable processes wherever life is abundant, life having developed from complex inorganic molecules that formed by chance out of simpler compounds accumulated under the influence of random events in certain local environments conducive to such formations, the environments and elements having been produced out of stellar trauma governed by the basic forces of nature, which appear to have disaggregated from simpler force(s) present in the earliest condition of the universe, which appears to have sprung spontaneously from nothing.

Or was it the other way around? Perhaps intelligence is the intrinsic reality. It is surely our means of experiencing everything else. Perhaps the separate packages of intelligence, arising and fading and moving about, are a consequence of our limited perception. Perhaps intelligence precedes the packages.

On the Red Line this morning I was struck by the role that our individual "packaging" of intelligence plays in keeping us apart. We carefully shuffle about as each additional passenger boards, maintaining quite precise, proportional spaces among us, and we become nervous if any passenger tries to narrow his separation from anyone else. Like the electro-weak force that acts to repel as well as to attract, the forms in which we assume consciousness to reside act to separate as well as to join us. Unless consciousness is organically localized by necessity, however, the potential to fill the gaps exists.

Perhaps the form of consciousness need not be organic, nor the packages forever discrete. A carbon-based intelligence is our only experience, but to assume it the only possibility seems an anthropic conceit. We once assumed an earth-centered universe, and still dream of an anthropomorphic God. Such hypotheses are not unreasonable in the absence of better data, but we do well to abandon them as contrary evidence accumulates. Our resistance to other hypotheses derives from our limited experience, and from fear.

The idea of non-localized or non-carbon-based intelligence seems at first fantastic, then offensive, then frightening. Consider, though, the fantastic sequence of unlikely random events that appear to have evolved the organic brain. Shall we be offended by a different sequence with a different result? Shall we be frightened by our own growing capacity to produce "artificial" intelligence, and frantically seek distinctions between it and the "real thing"? What is the real thing? Perhaps it includes many forms. Perhaps it offers connection, and extension, in time and space.

* * *

FRANCIS CRICK, a theoretical physicist, designed magnetic mines for the British Admiralty during World War II. He moved to the Cavendish group in Cambridge, where he met the young James Watson. It is said he was inspired by a book by Erwin Schrödinger, *What is Life?* [4] Schrödinger was also a theoretical physicist, from the preceding generation, a pioneer in the study of wave mechanics. He struggled with the implications of relativity and quantum mechanics, together with the great physicists of the first half of the 20th century.

In 1943, with the Second World War still in full contention and its outcome uncertain, Schrödinger undertook a series of public lectures at Trinity College in Dublin. His purpose was to explore how the complex processes within living organisms might someday be explicated by physics and chemistry, given their inability to do so at the time of his lectures. But in the process of the lectures (or at least of the book that he took from them in 1944), he explores the statistical nature of physics vs. the determinative action of reproductive life, and goes on to wonder at the "permanence" of inherited genes. During that wondering, he introduces the struggle of life against entropy, a theme that would become very popular in the "science and philosophy" writing of the second half of the century. The young Crick, reading Schrödinger's book, resolved to contribute an answer, and of course launched a revolution. Schrödinger's speculations on self-replicating "aperiodic crystals" may seem naive in the hindsight of Crick and Watson's double helix, but he derived the concept directly from physical principles before their data were available.

Schrödinger holds that time cannot obliterate Mind; that Mind is a singular reality of which (what we perceive as) individual consciousness is a part; and that consciousness is characterized by its recognition of novel (not repeated) events. This line of thought, like Marsalis's, recalls Shannon's information theory, in which the content of a message is proportional to its unpredictability.

* * *

THE HUMAN BRAIN is a great integrator. It has the capacity—through all the sensory organs serving it as well as in its processing of stored data—to examine great volumes of random events and perceive within them messages, meanings, forms, and connections and coherences that are deeply imbedded therein. It can create orderly patterns out of pure noise as well as it can perceive intended patterns. This is the source of beauty, says Bronowski; without such intelligence, every sunset would be only the inevitable product of random processes. Such pattern recognition, based on integration and correlation techniques to extract signals from otherwise overwhelming noise, is familiar in communication, in radio telescopes, and in such closer experiences as the human eye detecting slight motions and special patterns against a nearly infinite background of information.

The history of civilization is a history of integration. We have the time and leisure to meet today—indeed to communicate or speculate at all—because a thousand others are supporting us, feeding, sheltering, defending if necessary, teaching, helping, maintaining commerce and the general order. They also study the sciences that undergird our speculations, write of their own searches and insights, and make the arts that help us see beauty. The size and general organization of the human brain have changed very little in a million years. That it can cure disease and write symphonies owes to our learning to work together, not to our development of superior individual capacities.

* * *

WE CAN BE confident that our ability to manipulate another's brain (or our own), by psychological, chemical, electrical, or genetic means, will continue to grow. All of the perceptions and experiences of future humans may be quite controllable, and indeed we may be able to manufacture brains quite as sophisticated as those that have evolved "naturally."

We know also that our ability to share perceptions and information processing will continue to grow, as we accelerate the speed and capacity of chemical, biological, and electromagnetic networks and learn to use quantum-level computation and communication. Probably the development of more effective intelligence will exploit the intrinsic advantage of parallel processing, in which many individuals and "computers" work together toward common ends, supported by wide channels of communication.

Intelligent life is detected by its freedom and capacity to choose order and to communicate. We can build a processor that will discern, in limited cases, order from disorder, can exercise some choice based on its programming and heuristic processing, and can send, receive, and process limited messages; we would consider it to be imbued with intelligent life only if its freedom of choice in formulating and responding to messages could transcend limits that we gave it.

Are we to shrink from the "brave new world" of unfamiliar consciousness because it threatens our concept of a soul or violates what we have held to be holy? If we are to defend a sacred territory, it had better be bounded by concepts that are defensible on lasting grounds, not dependent on forms that are merely familiar. One of these concepts is our commitment to truth, wherever it leads. Another is our commitment to beneficial connection—to the understanding, preservation and benefit of our universe as we find it, including other forms of conscious life (those like ourselves and those unlike.)

A third concept, which derives logically from the first two, is the cherishing of freedom. To diminish another's capacity clearly works against our commitments; so would diminishing another's freedom, for both are included in our understanding of the expression of conscious life. Thus one might support synthesizing a new brain but not manipulating another's, unless the consequent loss of freedom is more than compensated by the increase in capacity—and that is a very hard trade to evaluate.

* * *

WE HOPE FOR a more universal intelligence, a "why" of creation, a central order, a ground of our being, in or through which we may increasingly connect with reality in experience and understanding. I suggest we seek this transcendent truth by being faithful as well as by direct search, because a rational faith means a commitment to truth and to each other.

Neither our faithfulness, nor our rationality, nor our commitments, would be violated by welcoming new forms of conscious life, whether we find them elsewhere or build them ourselves. We must apply the hard tests of truth and benefit—and freedom—to every new step we contemplate. Thus we rejoice in diversity, not as a matter of obligation but of freedom. We may eschew creeds but we can join in covenants freely, acknowledging our differences but seeking the greater good of our common enterprise. Diversity presents opportunity for new directions of growth, if we choose them. In the universe are likely many forms of intelligence, offering rich potential for connection, for growth beyond our borders to Marsalis's "edge of the possible."

We wish to find—that is, to understand and to experience—the ways in which life is shared. Our commitment to life requires us to reach out to other life forms in beneficial ways. Might we find a singular consciousness, as suggested by Schrödinger? [4] A central order, as hypothesized by Heisenberg? [5] Is all life one, as proposed by Paul Davies and others? [6] Our task is not to work a miracle but to understand and embrace what is already so.

As a working hypothesis, one might take this view, that a natural order is more likely to involve reaching out, learning more, preserving our environment, and drawing closer together, rather than separating and turning inward. It is the more compelling choice both for its consistency with what we know of the natural universe and its likelihood of finding some meaning or purpose in all that is. A purpose of nature should most sensibly be sought in its preservation, manifestation, contemplation, and discovery. For those who seek a purpose in life but cannot accept an established dogma, this is a sustainable faith.

* * *

IN SEEKING MEANING and a consistent basis of commitment, the scientific method should be as reliable as it is in dispelling the mysteries of the physical universe. Every cherished belief should be subject to the same rational tests that an unbiased scientist would use in challenging an established theory. Nor should we fix our world view on existing assumptions about the universe, lest they change and leave us groundless. That has been the fate of many religions, caught defending an ever-shrinking world not yet displaced by advancing knowledge.

To be able to trust the surprising, Shannon implies, is to experience truth. We must be able to test to accept; thus we cannot discern all of truth, because we have to correlate what we find with something else, necessarily leaning on some redundancy. Because we can correlate and share what we observe, however, our knowledge and skills grow geometrically. The accumulation builds, and the rate of building builds. You will solve many of the puzzles that have eluded the best minds on this stage today.

To those who protest that some mysteries should be left untouched, that science cannot understand love or beauty, we must respond that those things that we find lovely and beautiful should be enhanced by greater understanding, not diminished. All true experience lies within truth.

And how shall we define "truth"? Our definition will necessarily be imperfect, because we are inside the question looking out; moreover, we probably consider truth to be infinite, and to define is to limit. Nevertheless, if we are to use the word we should be clear about our meaning. I have suggested that we consider truth to be that which prevails in every event, which passes every test, which returns to every question an answer that prevails in every event and passes every test—in a mathematical sense, the integration of all information.

* * *

BUT WILL YOU continue to seek? Will you try to reach other worlds, or sink back in the comfort you have won? Will you fail to approach the edge of the possible, Marsalis would ask, for fear of losing balance? What will guide your decisions on applying and sharing your knowledge and skills?

The prospect of chemically- and electrically-induced happiness without harmful side effects is no longer a matter of pure imagination. Besides the widespread demand for such agents, there is the powerful economic pressure exerted by their manufacturers and distributors. One need only compute the per capita sales of unnecessary drugs in the U.S. to be in awe of the demand that has been created. Beyond these developments lies the prospect of slowing or even halting the process of individual aging. Huxley's *Brave New World* has already lost much of its satirical edge.

We have come so far because we have chosen to join and to wonder. When individual comfort can be commanded, will you withdraw and choose comfortable isolation? Our old religions gave the people not only courage to persist in the face of pain and certain death. They also called us to be together and to face pain for the benefit of others, present and future. Of course they also gave us excuses to quarrel and to be proud and to imagine differences and to resist learning sometimes. But perhaps we can forgive their imperfections and appreciate what they gave to our sense of beauty and our drawing together. Now we should turn from them with respect, as we take on more rational (and more faithful) principles.

We must, in fact, rely on truth to support our concept of life, to strengthen our faith, to bring us love. The progress of understanding our universe dispels not the refuge of faith but the clouds of ignorance. The claim of love can succeed only where supported by truth. Only that which survives the test of open inquiry can prevail.

We seek more of that reality. In particular, we should seek that which is lasting and universal, beyond that which is transient and personal. We need not eschew the latter to seek the former, but begin where we find ourselves and look outward: to seek what else, not necessarily what instead.

* * *

So, MY YOUNG friends (I would like to have said), we hold up to your study an unexplored potential, and to your contemplation a wonder. Pursue them with commitment and honesty, and they will give you meaning. Take joy in the deeper experiences of present being, and joy in the opportunity to seek more. You have the tools. Use them to raise the place on which others will stand.

Notes

[1] Jonathan Allen, ed., *Scientists, Students, and Society* (Cambridge, MA: MIT Press, 1970.) (BIM is a group of theologians, business managers, and scientists urging the R&D industry toward a rational consideration of the impact of their decisions on peace and justice.)

[2] Wynton Marsalis, *Making the Music*, WGBH/NPR, April 6, 1996. (See the letter from the Radio Communications Corporation for Shannon's theory.)

[3] Jacob Bronowski, *Science and Human Values*, rev. ed. (Harper & Row, 1965.)

[4] Erwin Schrödinger, *What Is Life?* (Cambridge University Press, 1967 [first published 1944].)

[5] Werner Heisenberg, *Physics and Beyond: Encounters and Conversations* (New York: Harper & Row, 1971.)

[6] Paul Davies, *God and the New Physics* (New York: Simon & Schuster, 1983.)

Gloucester Cathedral

ONCE TALL WATERS stood above this plain, and in them creatures not so far from us, who died to make our columns and our walls. In time the heave of continents thrust their bequest into the sun and the English rains stole its sandy cloak. The ancient life was raised again, in great limestone columns, supporting the Norman faith.

Now it is the winter of 1534, and King Henry VIII's accountants have assembled at the base of the great columns of Gloucester Cathedral. They have come here to complete their inventory of monastic properties, and it is an ominous visit. In four years, their labor will be complete, Parliament will endorse Henry's title to all church properties, and the great transfer of wealth from religious to secular control will change the western world. By then the Papal bull deposing Henry and absolving his subjects from allegiance will have been promulgated, but it will be too late. Economics will prevail.

From all the dissolved monastic churches, Henry will found just five cathedrals, including Gloucester in 1541. This will save for posterity a magnificent edifice and an ancient tradition, stretching back before Roman times. It will save Great Peter, the 14th-century bell still striking the hours today. It will save the A.D. 1350 stained-glass East Window, largest in England and perhaps in all of Europe, and the fan-tracery of the Great Cloister's vaulting, and the delicate stone work at the top of the towers. Your mom and I have climbed all over these miracles and wondered at the dedication of so many lives and skills to their raising.

But for me, it is the Normans' limestone columns that evoke the greatest awe. These simple, heavy columns stand where the people stood, here in the nave, separated from the cloisters, screened off from the choir and presbytery. Here the people came, gathered in hope of comfort, of meaning, of something beyond the pain and toil and shortness of their lives, of something that lasts a thousand years, that one could count on. I can feel their pain and longing as I stand here. The limestone bears visible evidence of fire and water, but I see deeper stains beneath its surface—stains from work-soiled hands reaching for a why, from blood of young lives lost too early, from tears of despair.

* * *

THE CATHEDRAL RISES above the plain even today. I can see it at a fair distance when I run along the English hedgerows above the town, and in medieval times it would have towered above the ordinary, reaching to heaven, standing plumb and square against the demons of the day, holding the secrets of the universal church. It was a time of magic when these columns went up, and faith a matter of believing in an organized, packaged, familiar, universal, super-natural hierarchy, arrayed against a capricious universe.

It was a dark time, a brutal, dirty, ignorant, crude time, a time of disease and cruelty and fear. We cannot today imagine what living then was like. Europe was gradually climbing up from the depths of the 14th and 15th centuries, when its population had reached its lowest point and little hope could be placed in any source, human or divine. Barbara Tuchman holds up *A Distant Mirror* for us:

> Associated with the cult of death was the expected end of the world. The pessimism of the 14th century grew in the 15th to the belief that man was becoming worse, an indication of the approaching end. As described in one French treatise, a sign of

this decline was the congealing of charity in human hearts, indicating that the human soul was aging and that the flame of love which used to warm mankind was sinking low and would soon go out. [1]

And William Manchester strains to see for us *A World Lit Only by Fire*:

As medieval men, crippled by ten centuries of immobility, [Europeans] viewed the world through distorted prisms peculiar to their age.

In all that time nothing of real consequence had either improved or declined. Except for the introduction of waterwheels in the 800s and windmills in the late 1100s, there had been no inventions of significance. No startling new ideas had appeared, no new territories outside Europe had been explored.... The Church was indivisible, the afterlife a certainty; all knowledge was already known. And nothing would ever change.

Shackled in ignorance, disciplined by fear, and sheathed in superstition, they trudged into the sixteenth century in the clumsy, hunched, pigeon-towed gait of rickets victims, their vacant faces, pocked by smallpox, turned blindly toward the future they thought they knew.... [2]

Yet in such a time magnificent cathedrals were raised, and Thomas Tallis could compose and direct a new music that would break out of medieval chanting as startlingly as navigation and science were breaking out of their primitive world-views, setting English choirs just as resolutely on a new course. [3] Now look what we can do when we sing together, Thomas, and shame, shame upon us when we fail to do it. When you composed, the world's pain carried the innocence of ignorance. Today much of it is of our own making. Which is the darker age?

It is in our hands to reach out to each other and to be one, to comfort, to share, to find meaning and purpose; but it can be done only together. We have learned to fly, Thomas, but not advanced your ability to soar above division, to harmonize. Is it beyond our reach to match technological with moral advances? Should this not be the objective of a contemporary faith? We stand beneath the great Norman columns and look about. Indeed this house of worship has lasted a thousand years, as its people hoped. Can we find our answer here?

The crowds have dwindled at Gloucester Cathedral, drifted away from mosques and temples, ignored the Arlington Street Church and the First Parish, lost hope or gone elsewhere for their comfort and meaning. And this is true not only of the selfish; I do not find many of our most thoughtful and caring friends in these houses of worship.

Yet we yearn for the transcendent. The longing, the hope, the search for a reality beyond our narrow selves—that cry echoing from this vaulted ceiling still springs from our modern hearts. All our comforts do not bring us quiet, nor all our winnings victory. Where is our faith?

In much of the world, the transcendent takes the form of gods or ancestors or an inclusive flow of life. Often an intermediary or personal savior is involved, as in much of Christian tradition. Such beliefs seem unlikely to many seeking truth in a scientific age, but our inability to derive comfort from the faith of others does not make them wrong or us right. Rather it shows a deeper need for a faith rooted in truth and in each other, a faith that may be more universally shared.

The need for meaning, for purpose, for being someone, is as strong as the need for comfort. We seek a basis for moral decision, some reality beyond our narrow vision, some assurance that there is, after all, a prevailing truth, that what we do matters, that love transcends our small estate.

To many of those who do not come through these doors, the evidence of a "God" as traditionally defined is not compelling. It is not self-evident that the scriptures or any other ancient texts are a matter of divine revelation. The churches and mosques and temples of our day seem built on anachronistic theories and writings that were beautiful and thoughtful in their time but ignorant of what we now know of nature.

Moreover, it seems to many that the God they were taught to believe in allows vast injustice to prevail. Neither justice nor compassion appears reliably built into creation or reliably invoked by their prayers. And when one looks across history—including very recent events—one sees the most monstrous atrocities committed in the name of those very religions that have enshrined love and brotherhood.

Of course there are answers to all these concerns, and of course our organized religions do great good and may be essential to civilization. They have lost, however, the consistent attention and commitment of many thoughtful, caring, fully informed, and rational people.

Are we better advised to keep our own counsel then, to be realistic rather than hopeful for a mystical salvation or afterlife, to be agnostic rather than baselessly faithful, to make the most of our own opportunities rather than to follow some archaic set of rules and dogma? Or can we find a faith that is based on truth and can be widely shared?

* * *

FIRST WE MUST ask what we mean by faith in an age without magic: how shall we define "faith" for ourselves and our children?

After all these millennia of wonder and creed, of wars and reconciliation, of despair and inspiration, all attributed to faith or the lack of it, can we at least agree on what it is? Perhaps we can agree that faith is a matter of commitment and trust; that to be faithful is to maintain our commitment in the face of adversity, trusting in this as the best course. (In this context, I include "hope" in the definition of "trust.")

This combination of commitment and trust is of course what we have said about love; and this is appropriate, since one should have a sense of love toward the object of one's faith. Then perhaps the "extension" of one's self in another, also associated with love, may be included in our understanding of faith. A strong faith might support an experience akin to the extension of one's identity. To be faithful would include an identification with the concept or being to which one is committed, whom one trusts. Such an extension toward the object of one's faith would strengthen the expression of "relation" in one's identity and might constitute the "becoming" we seek.

And what of rationality? What makes a faith "rational"? Like time and order, we all know what it means until challenged to define it. What do we mean by being rational? First we say it means being intellectual, as opposed to emotional. Perhaps we mean scientific, logical. Or perhaps just correct, not in error. Can an incorrect statement be rational? An illogical statement? An emotional, non-scientific statement? The dictionary appeals primarily to "reason," the capability of intelligent and (it is implied) dispassionate or unbiased processing of information, and the perception of reality by direct observation or deduction. It is not the same as being correct; one can be rational and wrong, or irrational and right. It is not the same as being devoid of emotion; one can be emotional and rational, or unemotional and irrational.

I suggest that the difference is intention. To be rational is to be committed to truth. And if we take truth as an object of our faith, then to be faithful is also to be committed to truth, since faith is a combination of commitment and trust. Therefore rationality and faithfulness are not only compatible; rationality *is* faithfulness, to truth. Such a faith is free of dogma and open to all advances in our understanding of reality. It is an unassailable faith, based in truth, fully communicable and lasting.

If my commitments are irrational I have no way of communicating or sharing them, beyond that small community of persons who will accept them without question. I aspire rather to a faith that could be universal, and such a faith demands sharing. The only medium I know for dependable sharing is truthful communication. A universal faith must be understandable by all.

My commitment is to rationality precisely because I take that to be most faithful. Anything less is intrinsically limited. This position does not disrespect the sharing of beliefs and traditions by the religious, for two reasons: first, they may be entirely rational, if they believe what they worship is true and are willing to put it to the test; and second, they evidently share a world view that provides a context for their communication among themselves (and usually their assistance to others), a common and comforting conviction.

If one believes that his object of faith is supported by truth when subjected to free and unbiased inquiry, and if he remains open to challenge and test, then his faith is rational. If he accepts the object without question and will not consider an argument, then his faith is irrational. One should not judge this as wrong, however, unless one can show the object to be untrue without ambiguity. Nor should one judge it inferior to one's own faith with respect to either truth or benefit, because one's own tests may be flawed, and the irrational faith may support great benefit to the faithful and those they touch.

Having admitted all this, however, my own search must be rational, and I believe there are many with me. For us, science and religion are not antithetical; in fact, to be faithful is to be rational. On that basis we are confident about a commitment to, and trust in, truth and each other. If an objective of a contemporary faith is indeed to match moral with technological advances, such a commitment and trust should help us meet it.

* * *

ARE TRUTH AND each other, then, the logical "objects" of a rational faith? I might be faithful only to myself; or my faith may stretch to others, to a transcendent being, or to a concept. When the object of my faith is another, a friend or family or tribe or community, we call that love: commitment to and trust in and extension to another. When the object of faith is a transcendent being or an abstract concept, we call it religion. The Catholic credo, still sung in this cathedral, is in a Trinity, while the more liberal Judeo-Christian-Islamic tradition trusts in a singular, all-powerful Creator. Mary Baker Eddy's definition of God was distinctly conceptual: Principle, Mind, Soul, Spirit, Life, Truth, and Love. [4] Other religions look to meditative enlightenment, reincarnation, and multiple objects of worship in the continuity and sharing of life. Comprehensive as this collection seems, it does not contain an answer satisfactory to many seeking a rational object of their ultimate commitment and trust.

I have argued that a rational faith would have truth as an object. This does not depend on beliefs that are not shared generally. Because it could be held universally and because it enhances intelligent life, we might agree that this is the highest faith. It is also logical to place our faith in something that we can trust always to be true, and the one abstract reality that meets that test is truth itself. Of course even the most logical among us will not always agree on what is true, but we can share a commitment to whatever proves reliable in every event, whatever meets every test. An abstraction of this concept—truth itself—may be considered without limit and worthy of our final trust.

To survive and prosper, however, and to meet the objective set in our encounter with Thomas Tallis, we need also a commitment to—and trust in, and identification with—each other. Michio Kaku speculates on why we have not heard from other civilizations in the teeming cosmos, pointing out the considerable probability that those who reach the skill to aim nuclear fission turn it on each other and destroy themselves before they reach the next level of maturity:

> It seems likely, therefore, that advanced civilizations sprang up on numerous occasions within our galaxy, but that few of them negotiated the uranium barrier, especially if their technology outpaced their social development. [5]

To prosper long in cosmological time requires far better cooperation than we have yet demonstrated. Like developing adolescents, our maturity lags our capacity. Trusting each other, committing to mutual benefit, we have no limits. Failing this, we have no future. The commitment and extension to each other fulfills life and enhances it. It is the rational road to salvation.

By "each other," a rational view of the universe would not end with the collection of *homo sapiens* on planet earth, but extend to intelligent life in all its forms. As an object of faith, Each Other must be taken in its fullest sense. I will call this the "Whole Each Other," which invokes both senses of the word "whole"—as inclusive of all, and as integrated, complete.

Martin Buber's sense of the "thou" included others as an extension of self, and was not limited to the few individuals who are the objects of our private loves:

> Extended, the lines of relationships intersect in the eternal Thou. Every single thou is a glimpse of that. [6]

Buber is speaking of a transcendent reality that many call "God." To what extent does a rational faith support such a hypothesis? In other words, if we grant that truth and the whole each other are reasonable objects of a rational faith, does reason support their intrinsic reality, transcendent of local instances? Is there "something there," to which/whom one might address one's commitment, beyond an abstract concept of truth and the benefit of others? Roger Penrose, supporting the notion of objective reality in basic mathematical principles and counter-intuitive quantum-mechanical phenomena, testifies that:

> I have made no secret of the fact that my sympathies lie strongly with the Platonistic view that mathematical truth is absolute, external, and eternal, and not based on man-made criteria; and that mathematical objects have a timeless existence of their own, not dependent on human society nor on particular physical objects. [7]

Shall we extend this faith in objective truth, and in a whole each other, to a transcendent absolute? (Here I do not mean to personify our search, only to establish a reality at a higher level, much as Penrose subscribes to truth existing without dependence on human observation, or as we accept an intrinsic existence of symphonies beyond their assembly of notes, of machines beyond their parts, etc. Nor do I mean mystical or supernatural when I say "transcendent." I am simply looking for the same kind of objective reality in our concepts of life and truth that we find in any whole beyond its components.) A commitment to this "Whole Each Other," with whom we share the capacity and freedom to organize and communicate, would be a faith in a life potential whose manifestations we cannot limit to our own experience. We might call such a potential universal. If I may use religious terminology for a moment (since we are standing in Gloucester Cathedral), I would say that a universal intelligence (or at least its potential) would be a "God" whom the rational could trust.

* * *

WHEN WE SEEK life elsewhere in the universe, we look for signs of intentional order and of communication; it is our only clue. As far as we know, each separate expression of life eventually decays. But the potential for life—for the great improbabilities of consciousness and freedom sustained in special places against Nature's downward slope—this potential exists generally. Just as we can speak only of the potential for elemental particles to exist at a certain point with a certain momentum, we cannot tell where intelligent life will spring up, or what form it may take. But we know it can be.

Reasoning upward, so to speak, from a rational view of life and identity, one should expect an absolute reality to be characterized by the completion or perfection of those capacities we have hypothesized as evidences of life, and of those qualities associated with identity. Thus one should look for the ultimate extension of order, freedom, and communication (quantitatively); and of commitment, relation, and transparency (qualitatively.) Taken together, these could be considered an ultimate projection of our concepts of life and of identity.

How might we stretch each of these components toward such a completion? "Order" might be projected on an inclusive organization or "Central Order" as suggested by Heisenberg [8]; "Freedom" on that which is entirely without limit, thus Whole; "Communication" on the holding of all information, thus Truth; "Commitment" on the unconditional; "Relation" on the universal; and "Transparency" on the perfection of clarity, devoid of any opacity to reality.

These projections suggest the nature of a transcendent reality based on our present experience at the "individual" level, at this point in our evolution. Of course the ability to promote this experience conceptually to a higher level does not prove it is "there." The existence of such an objective reality, however, appears to me more likely than its absence, much as the intrinsic existence of mathematical concepts seems more likely than their limitation to particular applications. In any event, it is the apparent *direction* of the evolutionary process—likely in other life forms as in ours—which is a history of increasing complexity, organization, integration, freedom, and communication. I suggest that this point in our human evolution is not a completion but a moment in becoming.

Perhaps we can trust this potential and commit to it, as to the central order speculated by Heisenberg, to an integrative reality that calls us to each other. It is not a bond we seek, but a free joining to mutual advantage:

> The problem of values...concerns the compass by which we must steer our ship if we are to set a true course through life....I have the clear impression that all such formulations [of philosophies] try to express man's relatedness to a central order....But in the final analysis, the central order, or the 'one' as it used to be called and with which we commune in the language of religion, must win out. And when people search for values, they are probably searching for the kind of actions that are in harmony with the central order... It is in this context that my idea of truth impinges on the reality of religious experience. I feel that this link has become much more obvious since we have understood quantum theory... [8]

Heisenberg understood "order" as well as anyone, and his sense of a "central order" as a moral compass is based on a rational concept: that self-interest can be found in commitment to a common good, to cooperation in building, to becoming together. The capacity to seek this symbiosis, and the potential for a higher order, confront us everywhere; and if the potential for life exists beyond our limited experience (which certainly seems likely), that capacity and that potential for reaching out extends beyond our present grasp.

* * *

THERE IS NO containing truth. It is not owned by any religion or captured by any preacher, nor is it defined by words or entirely found by seeking. One should not be defeated therefore by realizing that truth may not be contained, captured, defined, or entirely found. It can be sought, and the search is fulfilling in itself. It needs neither auspicious beginning to be legitimate nor glorious end to be successful.

One should expect that the advance of science will change our questions as well as how we try to answer them. That which expands our understanding of what we observe and how we observe it necessarily affects the context in which our most basic search is conducted. We have only our personal and shared experience to go by, whether we call it knowledge or revelation. Therefore we should happily accept any contribution to our insights. When we look for reality, we should raise our sights. One is more likely to perceive truth in a view that cannot be fully defined than in one limited to the familiar.

In 1987, Stephen Hawking concluded a lecture on "The Origin of the Universe" by speculating that the laws of physics, as we know them, may not break down even in "singularities" such as the beginning of the universe:

> What is it that breathes fire into the equations and makes a universe for them to govern? Is the ultimate unified theory so compelling that it brings about

its own existence? Although science may solve the
problem of how the universe began, it cannot answer
the question: Why does the universe bother to exist?
I don't know the answer to that. [9]

No one does. [10] Moreover, there are other "whys," and there are
some hypotheses that might be advanced toward theories to satisfy
them. Why should freedom develop in a universe that seems, in
nearly all its vast extension toward both the huge and the microscopic, so devoted to the Second Law? Why is the universe not
capricious; i.e., why does truth prevail in every event? Why intelligence anyway? For that matter, why are Hawking and Gould and
others who find no evidence of a transcendent purpose so committed
to truth and to the preservation of life?

One can adopt a statistical and reductionist position that
is internally consistent on such questions, and can assume that
any meaning of life—including a commitment to truth and each
other—is put there by ourselves. That is, our very special universe
or statistical history is one of an infinite number of possibilities, the
one that allows us to be here and ask the question; brains evolve
by natural selection and accident; and once evolved, can imbue life
with meaning through their own behavior. This is OK. I would not
impose God on the agnostic any more than I would steal God from
the believer, and all of us share commitments to each other.

A more cogent set of hypotheses, however, emerges from the
deep encounters of which I write to you—from our great scientific
advances, the study of our religions and philosophies, the cries of
those who struggle and the response of those who help, from all
these as they build a shared and coherent vision—a set of hypotheses
that is internally consistent, and the more compelling for its sharing.
My reading of such hypotheses would include these:

There is something rather than nothing. Truth prevails in every event. It may not be possible for us to prove an "objective reality," as opposed to a set of observations and laws that simply explain our experiences, but the whole of truth exists independently of our observation.

Although the universe is probabilistic, it is not capricious. We have the ability to assess probabilities. Within the apparent limits of Heisenberg's Uncertainty Principle, the laws of physics are dependable and seem universal. We cannot pin things down precisely because we are part of the reality we would define, but we can make dependable predictions of effects from understandable causes.

The universe (or the collection of universes or alternative histories) is not entirely deterministic. Within the general trend toward equilibrium and increasing entropy, intelligent beings exist who have the capacity and the freedom to alter the probability of future events.

This capacity and freedom to increase order locally, although it comes at the expense of greater increases in the general disorder, allows building, organizing, and modifying the environment for mutual benefit. It is unlikely that any achievement is ultimately beyond our reach, working together. (Obviously the same capabilities can destroy the environment and its residents.)

Among the capacities of intelligent beings is the capability (and the proclivity) to exchange information concerning their respective conditions, environments, and intentions. Such communications, which in our form exist at the cellular as well as inter-personal levels, support the sustaining of conditions far from equilibrium, the great improbability we call life.

Conscious life is more likely to be widely distributed than to be limited to our local species, but it probably takes many forms, some of which may be so beyond our present experience that we shall have difficulty recognizing them.

The media and means of communication have probably not been exhausted. It is not likely that all forms of intelligent life will have chosen our media or our means.

Our view is unlikely to encompass all that can be.

Truth is more than the sum of facts, and intelligence more than the sum of brains.

The elimination of particular forms of conscious life is unlikely to eliminate the reality of intelligence. Intelligence is not necessarily limited to local, discrete, and temporary forms. Reason supports what Schrödinger called the "unlikely eradication of Mind by Time." [11]

Among intelligent beings, there is therefore an unexplored potential of connecting, of sharing.

Communication, as the sharing of information and reduction of disorder among intelligent beings, may provide access to what Heisenberg called a central order. [8]

Although the death (disintegration) of individual brains is evident, so is the life shared (integrated) by communicating intelligence.

An infinite number of courses of events is possible (with varying probabilities), and the universe may take many branches. We can affect these probabilities, in proportion to our communication with (a broadly-defined) each other. This suggests a possible connection with—or at least an evolution toward—a central order and an inclusive intelligence.

This is not mysticism, and it does not require an identification of a divine source—although it would not necessarily be inconsistent with either. This is a course based on the pervasiveness of truth, the reality of our sharing all things with the universe, and the freedom of each individual to seek connection. If we consider Truth and the Whole Each Other, most broadly construed, as the object of our faith, we have a way to live that is both rational and faithful.

I trust in a Whole Each Other because that seems to me the most compelling answer to the "why" questions. Others may have no such expectations, and I have no need to convince them. Whatever one's expectation of the outcome, however, the search for truth and the beneficial engagement of others gives us equal meaning, makes us equally faithful. I do wish others to make such a search and engagement—such a commitment—for it is fulfilling in itself, and we walk together.

Faith has always been a combination of trust, commitment, and extension. As a people matures and magic recedes from their understanding of the world, the old objects of faith withdraw from everyday experience. The realm of traditional religion retreats as our understanding advances. In the end, only truth prevails, and the capacity of life to seek connection. The faith of a mature people can be to trust and to commit to truth and beneficial connection, to our becoming together. For faith to last, to stand the test of time and the challenge of new discovery, it must be rational. Such a rational faith, rooted in truth wherever it leads yet mutually supportive, should be a source of comfort and of meaning for all time.

Notes

[1] Barbara W. Tuchman, *A Distant Mirror: The Calamitous 14th Century* (New York: Random House, 1978), 588.

[2] William Manchester, *A World Lit Only by Fire: The Medieval Mind and the Renaissance* (New York: Little, Brown, and Company, 1992), 26–27.

[3] Thomas Tallis, 1505–1585, served in the courts of Henry VIII, Edward VI, Mary Tudor, and Elizabeth I. Elizabeth granted him and Byrd, in 1575, one of the first exclusive patents to print and publish music.

[4] Mary Baker Eddy, *Science and Health* (Boston: Christian Science Publishing Society), 587.

[5] Michio Kaku, *Hyperspace: A Scientific Odyssey Through Parallel Universes, Time Warps, and the 10th Dimension* (New York: Anchor Books [Bantam Doubleday Dell], 1995), 289.

[6] Martin Buber, *I and Thou*, trans. Walter Kaufmann (New York: Scribner, 1970), 123.

[7] Roger Penrose, *The Emperor's New Mind* (New York: Penguin Books, 1991), 116.

[8] Werner Heisenberg, *Physics and Beyond: Encounters and Conversations* (New York: Harper & Row, 1971), 214–216.

[9] Stephen W. Hawking, "The Origin of the Universe" in *Black Holes and Baby Universes* (New York: Bantam Books, 1993.)

[10] I recognize that there are millions whose answer lies in their belief in a creative God. I hope they will forgive my inability to accept such a hypothesis without question.

[11] Erwin Schrödinger, *What Is Life?* (Cambridge, UK: Cambridge University Press, 1967.)

The Berkshire Hills

I T IS NOVEMBER and raining. Not enough for umbrellas, just sufficient to exorcise any warmth or brightness that might have crept into the late New England fall, just after color fled the land, just before the forgiving snow.

I look above the heads of the quiet crowd around me, at the empty branches nearby, receding darkly into the woods and climbing the misty slopes of the Berkshire Hills. Robert Frost's lines about November keep running through my mind: the desolate, deserted trees, the faded earth, the heavy sky...

I have forgotten something. The director gestures to me and I recall my final obligation: to remove my gray gloves and drop them on Winston's[N] casket as it descends into the damp ground. Here, then, are my gloves: my hands cannot reach you, were not a sufficient support, could not hold you up or thrust away the darkness that took your life.

How could this be? He was so bright, so promising, destined for large events and long reaches beyond this sad corner of earth. Why? Where did we fail? What should we have done? A cold rain falls through our own darkness, dissolving us all in the fresh grave.

Can we survive this despair? Can we comfort each other, find the meaning that eluded Winston? Does a civilization really advance as it solves the mysteries of life and learns to postpone death? In some ways we have come so far, yet truth still feels cold to touch, and we wonder how to replace the old religions with a rational faith that gives some warmth.

* * *

PASSING OLDER GRAVESTONES as we walk back, my mind drifts eastward and backward. It is 1775 in Bolton, Massachusetts, six months after neighbors from Acton drove English troops from an old bridge in Concord, and the Colonial revolt is still uncertain. Mary and Oliver Pollard do not care: they have come to bury little Jonas, three-and-a-half-years old, and baby Rebeckah, one year, one month, and twenty days old. They died on the same day, "Oct ye 5th," probably of a disease that is easily prevented today. Mary and Oliver, helpless, have watched their little ones suffer, cry for comfort that could not be given, and finally slip away.

Now they are listening to prayers to a God whose mysterious ways cannot be understood, trying to reassure six-year-old Stephen at their side. Stephen has reason to be troubled: not only have his two little siblings died together, but his older sisters Ann and Mary are lying ill at home, too weak to make the sad trip. Ann will die on the 9th and Mary on the 17th, having reached the ages of eleven and fifteen.

Oliver and Mary Pollard would have buried 15-year-old Mary, their first-born, about this time of year, after the color had been washed from the trees and torn from their lives. Perhaps it was raining that day too, the earth gone brown and gray, the air heavy. Perhaps they felt as we, but more drained, more defeated; or were parents then more hardened to death, always so close, so inexorable? Surely they were more fatalistic, having so little understanding and so little control.

Little Stephen will make it through the winter but die next May. Again and again Mary and Oliver will walk down this old road and watch their immortality slide into the earth, one little grave after another. They will try to bring back some of their lost joy by giving new birth, naming a new Stephen and a new Mary. One will die after "1 yr 1 mo 27 days" and one after "2 yrs 2 mos

3 days." Finally a second Jonas will be born and outlive his parents into the nineteenth century: one child out of eight. Somehow they will survive their unspeakable tragedy and live together for another thirty-seven years, dying within a year of each other as the young country's second war with the English begins. [1]

* * *

IN OUR OWN grief we wonder at their survival. No happiness seems possible now, no interests, no energy. We have no hope in a future. The color is gone, the flavor of life, the inspiration. The mockingbird has lost his voice, the raven caws his nevermore. Yet Mary and Oliver went on, and their second Jonas; they loved again, and laughed and sang and built, built for us.

They built for us. And now look, Mary and Oliver, what we could do together: smallpox is unknown in the world. Most of the demons that so haunted your world have been exorcised from ours. In ten generations we have raised on your foundation a home of promise, of possibility, nearly safe from the demons without.

Yet we are not free from the demons within. Our home is still primitive in many ways, still insular. We distrust intrusion still, and seek our comfort and meaning by drawing shelter and secrets about us. We slip easily into quarreling, but now with deadlier effect than you could imagine. Our view of the world, our expectations of nature, our notions of other life forms, are still anthropocentric. Still we wander.

* * *

IT IS A SMALL notion of life that can dissolve in a grave, a small sense of identity that can be cut off at such tight edges. We recognize intelligent life, we have said, by its ability to discern order, its freedom to choose, its capacity and intention to communicate; and we have said that we recognize identity through commitment and transparency and relation. A good question for those in a cold and rainy cemetery is, What level of individuality owns life and identity? Gould will not permit us to pin it down; Davies says we are all one. [2] We might then consider whether one's claim to the attributes of life and identity are enhanced or diminished by a wider perspective, by seeing life as a shared reality.

Our capacity to discern order, to bring it forth out of chaos, to build, to organize, to sustain the improbable, is not an "individual" capacity, in the sense of belonging exclusively to a particular level of biological hierarchy or evolution. It is a capacity continuously enhanced by cooperation, by joining. Nor can the "individual" communicate when alone, nor, I think, is she most free alone.

Freedom is unique to intelligent life. The crystal organizes spontaneously from a cooling liquid, the noisy cosmos sends forth enormous information in all directions, but only intelligent life demonstrates choice in its organizing and communicating, its freedom to intervene among probabilities, to affect its future. Is such freedom encountered only in the "individual," and are we sure that the isolated human organism is the only legitimate individual? The degrees of freedom, the range of choices enjoyed by increasingly complex organisms expand in general with the number and variety of interconnections possible in their make-up. Our freedom itself, however much we cherish individuality, grows in collaboration. Of course we can be constrained, even imprisoned, by each other, and sometimes wish to escape from unwanted company; but our choice of order is liberated and the barriers to it reduced by coherent integration of our endeavors. We shall be better at building, learning, and healing as we join in surmounting old limits and solving old puzzles.

Regarding the expression of identity, my range of commitments is narrowed in isolation, I cannot be transparent to them without others, and I cannot "stand in relation," as Martin Buber says, by myself. Thus it seems the attributes both of intelligent life and of identity are indeed expressed more fully with increasing integration.

Buber challenged us to advance from "I-It" relationships to "I-You" (Du) relationships. [3] In my visits to places of great striving over these years, I hear a challenge even more demanding than Buber's: to stretch the "I" toward the "we." Our struggle to retain an isolated, individual life, to own it within defined boundaries, is not just futile; it is self-defeating in the broadest sense, i.e., it defeats the broadest sense of self. It defeats the chance to share life, to transfer meaning, to widen identity, to be fully free.

* * *

WALKING THROUGH THE rain, I imagine a service different from the graveside finalities we have witnessed in the Berkshire and Bolton cemeteries. I invite Martin Buber to officiate as rabbi, and Mary and Oliver Pollard to bring their perspective on two hundred years' progress. I ask Heisenberg to speculate again on a central order and Schrödinger on his sense of a singular consciousness. I invite Bronowski to speak on the use of analogy and elegance in art and science, and how we make our discoveries of "hidden connections" into real advances. Claude Shannon must attend to help us consider the central role of information in a probabilistic world and of communication in stretching it; and of course Einstein must be here to lead us in some thought experiments, beyond intuitive limits. Finally I want my old friend Abigail Eliot to chair the service and to keep us from nonsense.

I try to imagine what direction of thought Abby's service might commend to us. Often our perceptions are limited by long-used concepts, ingrained ways of thinking, or the burden of our isolation. The great advances of art and science have frequently proceeded from leaps over such limitations. Too often, however, we passively wait, not perceptive or courageous enough to cast off conventional thinking or leap to the potential that may be found in a more pro-active reach.

Perhaps Einstein would set up this "thought experiment" for us: imagine that we do receive an apparently purposeful signal from Harvard's antenna (or from some other medium of communication), a signal that appears to bear evidence of extraterrestrial intelligence. How shall we proceed? We have two sets of questions to answer, one quantitative and one qualitative:

(1) Is this legitimate evidence of intelligent life; that is, to what extent:
  • can it distinguish order from disorder, and therefore have the capacity to organize, to build, to integrate, to civilize?
  • is it free to choose order, to determine its future?
  • can it communicate?

(2) What is the intent, the character of this life form? That is:
  • what are its commitments?
  • to what qualities and realities is it transparent for us?
  • what relationships does it have, can it make, does it intend?

Positive answers to the first questions establish only that intelligent life is "there," that we are not alone. This does not describe the nature of our neighbor: the first answers deal with the "what" questions, not the "who" questions. We would quickly want to know more about her morality, what he believes in. In fact, we are trying to establish what we have called an "identity," exactly as we establish our own. It is not enough to live; one must be.

* * *

Buber, of course, would remind us that we find ourselves in others, in relation. He would use the extraterrestrial visit to illustrate how universal that relation must be. I think he would suggest that the "transparency" element of identity also becomes clearer when hypothesizing an extraterrestrial intelligence: what windows open for us as we enter relation with this new "person"—what do we see more clearly than before? what do we learn? what new experience becomes possible? what wider perspective on ourselves opens up? what special beauty can we now discern? what new view of grace, strength, or courage now makes such possibilities real for us? Of course this is our terrestrial experience also—we just confuse the window with the view, the medium with the message.

Mary and Oliver would speak of their faith in the inextinguishable human capacity to love, a capacity not limited to particular individuals, shared by those who have lost or never had a loved one, accessible to all whose relationships do not meet their needs. They might call this the Love of God. But all of us have felt, in the presence of beauty or discovery or goodness, the tug of a transcendent caring that the universe seems to manifest. Can we make that feeling, that hope, real?

Love and faith reside in our capacity to commit and to trust, to seek shared context and to build relationships that extend our identities. Shannon would remind us that intelligent beings can build a shared context and strong relationships through truthful communications and open "apertures." That capacity should be as universal as the intelligence at its ground; so what can we say of that? Can we say that I will die and You will die but We will not?

Well, Abby would say, this is obvious at the level of genes and of species. Individual aggregations die but families and species go on. To me it is compelling also at the level we have called "identity": the "I" in isolation can die, and need not wait for death; he need only choose indifference over commitment, opacity over transparency, and isolation over relation. If instead he stretches toward the "we" in connection, intensively sharing, his identity broadens. If love is expressed in commitment (and trust) in relation to others, and if identity is expressed in commitment, transparency, and relation, then sharing love is sharing identity, sharing our selves. Such an investment in others seems to me a rational act of love, an act of rational love, finding self-interest in other-interest.

Abby would interrupt me at this point: are we drifting into mysticism and wishful thinking here, or are there rational grounds for your "we"?

Heisenberg might answer in the context of a "central order":

> Can you, or anyone else, reach the central order of things or events, whose existence seems beyond doubt, as directly as you can reach the soul of another human being?.... I would say yes...the word "soul" refers to the central order, to the inner core of a being whose outer manifestations may be highly diverse and pass our understanding. [4]

This suggests an unexplored potential of sharing. It would depend on communication among elements of intelligence, each with the freedom to choose joining or remaining aloof. Our closest bilateral sharing is our limited experience of love. A multilateral sharing would be more powerful and more consistent with our understanding of full living, of the meaning of life.

Einstein would elaborate on the concept of reaching and sharing by reminding us that the separation we experience from each other is an "optical delusion" of the individual consciousness, a "kind of prison for us," whereas "A human being is part of the whole..." [5]

We have not really explored this potential very far or very long. Erwin Schrödinger suggests we seek a more universal intelligence:

The only possible alternative [to "the invention of souls, as many as there are bodies"] is simply to keep to the immediate experience that consciousness is a singular of which the plural is unknown; that there *is* only one thing and that what seems to be a plurality is merely a series of different aspects of this one thing, produced by a deception.... I should say: the over-all number of minds is just one. I venture to call it indestructible since it has a peculiar timetable, namely mind is always now. There is really no before and after for mind. There is only a now that includes memories and expectations....we may, or so I believe, assert that physical theory in its present stage strongly suggests the indestructibility of Mind by Time. [6]

Abby asks if such speculations are not rigorous, unsuitable for testing, in spite of the scientific credentials of their authors, but Bronowski points out that analogy and metaphor are at the root of scientific advances:

To us, the analogies by which Kepler listened for the movement of the planets in the music of the spheres are farfetched. Yet are they more so than the wild leap by which Rutherford and Bohr in our own century found a model for the atom in, of all places, the planetary system?.... All science is the search for unity in hidden likenesses.... The progress of science is the discovery at each step of a new order which gives unity to what had long seemed unlike.... Science is nothing else than the search to discover unity in the wild variety of nature—or more exactly, in the variety of our experience. Poetry, painting, the arts

are the same search, in Coleridge's phrase, for unity
in variety. [7]

Where might we look for evidence of an abstract reality suggested
by the concepts of "mind" or "consciousness"; i.e., of intelligence not
conventionally limited or divided? We cannot devise a conclusive test
of such a hypothesis because we are contained in it. We run into the
same inherent limits that led Heisenberg to the Uncertainty Principle and Kurt Gödel to the fallibility of mathematical proofs. [8]

We can recognize, however, that the things that matter—
intelligent life, beauty, freedom, trust, commitment, truth, the
capacity to build, to learn, to heal, to communicate, to love—all
depend on information. All would therefore be enhanced by more
perfect communication and embraced by a "central order," a more
universal intelligence. Hanging on to the individual brain, to a set
of memory cells, is futile in a search for immortality, but seeking the
reality of inclusion in Intelligence is a rational reach. We would not
need a magical intervener, a tribal god of war, a personal savior, a
creator, an anthropomorphic supporter, a receiver of our souls, if we
could perceive such a reality.

The inability to devise immediate tests for cogent hypotheses
has not prevented their acceptance as working models (subject to
adjustment when better models are devised or new findings prove
them wrong.) In fact nothing can be proven beyond any residual
doubt, not even in "abstract" mathematics, as established by Gödel in
1931. [8] Many of the strange predictions of relativity and quantum
mechanics had to wait a generation for experimental techniques to
catch up. Yet the predictions, and the theory supporting them, were
accepted as worthy of pursuit, because the theory "fit": it provided
a cogent explanation of unexplained phenomena, it squared with
experience, and its mathematics were elegant.

We choose new hypotheses as our working models, even
if beyond present test, when (a) they offer credible explanation of

mysteries or contradictions; (b) they can be expressed in an elegant or beautiful form, i.e., a form that is efficient in revealing the order in apparent chaos; and (c) they disclose hidden connections or bring unity to variety. Scientific history shows this to be a reliable instinct, even when the new hypotheses raise new questions and lead to strange consequences well beyond our experience.

Michio Kaku, in his speculations on ten (or twenty-six) dimensions, concedes that the energy required to test his hypotheses cannot begin to be assembled in today's laboratories. The use of many dimensions, however, makes the mathematics of relativity and quantum mechanics consistent and simpler, and therefore they have to be taken seriously in spite of their counter-intuitive implications. Kaku says this has happened often before. When a new hypothesis meets tests (a)–(c) above, serious investigators will often base substantial work upon it and often make substantial progress that would not have been possible without it. By trusting the direction in which it points, even without full confidence in the implications or full understanding of the details, progress exceeds that which would have been made absent the speculation. [9]

Other examples are found in analogies to the concept of entropy in thermodynamics. The "Second Law," familiar to engineers and physicists since the nineteenth century, notes that disorder always increases in energy transfers (except those that are reversible, rarely encountered.) When Claude Shannon named his famous measure of information content "entropy," he was recognizing its mathematical equivalence to that concept. Subsequently it has been shown that the two measures of entropy (thermodynamic and informational) can be derived from each other.

Statistical mechanics shows the "disorder" of heat to be equivalent to our degree of ignorance concerning the complex motions of all the particles whose combined "random" (i.e., unknown) kinetics have been increased by the dissipation of a more ordered energy into heat. Then in 1970, mathematical modeling showed that a black hole's boundary always increases as additional matter falls in. Hawking and others recognized the analogy to the Second Law, and felt there might be a "connection" between the behavior of black holes and thermodynamics. This led to a successful theory of the entropy of a black hole. [10]

* * *

So IT MAY be with hypotheses on pervasive and intrinsic intelligence, a central order, a potential for sharing life, communication via unfamiliar media, and other means of stretching our long-held limits. We are frustrated at the moment by being inside the puzzle ourselves, and by our inability to devise a conclusive test. This need not inhibit our taking on the direction, however, stretching our means of communication, opening new possibilities through our unfulfilled capacity to work together.

The advance of our sciences and our ability to cooperate has bought us, over the millennia, the luxury of quiet. We can still the noise of pain and fear long enough to listen to farther signals. It seems the time has come to seek beyond the limits imposed by ancient fears. If we can be freed of limiting concepts and discarded world views, why not seek a broader identity? We are no longer children. We have the means to stretch.

Notes

[1] Mary and Oliver Pollard, gravestone carvings, cemetery at Bolton, Massachusetts, 1775–1812.

[2] Stephen Jay Gould and Paul Davies: see the letter from the Musée Marmottan.

[3] Martin Buber, *I and Thou*, trans. Walter Kaufmann (New York: Scribner, 1970), 123.

[4] Werner Heisenberg, *Physics and Beyond: Encounters and Conversations* (New York: Harper & Row, 1971), 214–216.

[5] Albert Einstein as quoted by Heinz Pagels: see the letter from the Metropolitan State Hospital.

[6] Erwin Schrödinger, *What Is Life?* (Cambridge, UK: Cambridge University Press, 1967); 94, 95, 145, 165.

[7] Jacob Bronowski, "The Creative Mind" in *Science and Human Values*, rev. ed. (Harper & Row, 1965), 12–20.

[8] For a reasonably accessible presentation of Gödel's Theorem, see Douglas R. Hofstadter, *Gödel, Escher, Bach: An Eternal Golden Braid* (New York: Vintage Books/Random House, 1989), 15–19.

[9] Michio Kaku, Hyperspace: *A Scientific Odyssey Through Parallel Universes, Time Warps, and the 10th Dimension* (New York: Anchor Books [Bantam Doubleday Dell], 1995), 185–190.

[10] Stephen W. Hawking, *Black Holes and Baby Universes* (New York: Bantam Books, 1993), 104–106.

Tilden Pond

I T IS A SONG of transience, a far echo, a ghost's call, the solil-oquy of a shadow. It is the erosion of time on the slopes of our repose: a sigh of resignation to an infinitesimal present, caught between a future that cannot be grasped and a past that cannot be held. Yet it is an assurance of possibility, a prescience of tomorrow.

One does not forget the call of a loon, alone on a lake in the night.

He is heard only out here. He has not moved into our sub-urban wetlands, as has the great blue heron, or proliferated around our puddles as the Canada goose, or invaded our gardens as the white-tail deer. He is wild still, elusive and mysterious, diving and swimming before the ice comes.

I listen on this porch every night, alone in the growing chill with the loon and less familiar sounds, writing, wondering. The loon sings his song of transience, and I wonder.

For example, I wonder about what makes a past and a future. I tear off the corner of today's page in my appointment book and let it drop to the rough floor boards, swinging and fluttering down for perhaps a fleeting second of something we call Time. The past is real only as memory, the present only as transient experience, in an infinitesimal slice of time. And the future?

Well, it is not Newtonian; that is, the future is not entirely determined by what has gone before, because of irreducible uncer-tainties in all initial conditions, their amplification in chaotic pro-cesses, and the continuing uncertainty in random events (not to mention the peculiar intervention of intelligent choice.) The universe takes paths we cannot know with certainty and cannot affect entirely. The future is therefore not predictable, and does not exist in our slice of experience. But the capacities of intelligence and its freedom do give us something to say about the future, permit us to alter probabilities, allow us to make use of the "remembered information" that is our past.

Information supports life, and it conveys truth. It is the heart of beauty, that which is conveyed in communication, that which defines order. Information theory says, however, that only new information has measurable value because its content is proportional to its uncertainty; once received, it is no longer informative. (In Shannon's equation, the logarithm of the probability of every received bit becomes zero, because its identity is now certain.) This is why we publish and broadcast the "news," why we speak of "good news" and "bad news."

Not all "remembered information" is useful. Some, however, gives us the context essential to receiving further new information (for example, the alphabet, the channel to be used in our receiver); it is what teachers convey, what gives meaning to our art and foundation to our building. Our context includes all of any message preceding that part currently being received; and, since the present is infinitesimal, essentially all received information is "remembered." [1]

Traveling in the Southwest, your mom and I pass innumerable oil and gas wells, marked only by modest pumping stations and pipelines. It occurs to me that these should be considered "information assets." The visible assets are only a small part of the total value they represent: finding the deposit and drilling into it were the major investments. Now all that uncertainty and speculation and searching has yielded to the obvious: the location of the deposit is no longer new information—the information is "remembered," and that is where one pumps.

This is the essence, the value and reality of the past: remembered information. It exists in all systems with memory. Whereas new information lives only during the instant of recognition, remembered information lives as long as the memory (often collective and augmented) in which it is stored. It may support our organization (as in construction practices), our freedom (as in maps and keys), and our context for communicating (as in alphabets and common experience); thus it is essential to the continuing of life. Or it may be of zero or negative value, if it inhibits organization, freedom, or communication—for example, lies and trivia.

If this is the past, what of the present? Can one live in an infinitesimal slice of time? Returning to Boston, I carry the puzzle to a road race, where time is everything. This race starts at the Museum School, across from and sponsored by Boston's Museum of Fine Arts. The Museum School is familiar to me; it is the source of our Prison Art Program interns, teaching the State's most serious felons at Walpole Correctional Institution, trying to get them to express their anger and frustration more in art and less in behavior. The interns are young and idealistic, having not yet learned what is impossible, and therefore willing to try it. Occasionally they succeed. This amazes our patrons and board members, whose spheres of interest and influence had seemed so inclusive before their introduction to a different and far-removed "real world."

Maybe what makes a real world—a term my venture capitalist friends like to invoke when we entrepreneurs promise too much or deliver too little—is the same as what makes an infinitesimal present real. Maybe it has to do with becoming instead of arriving. I will think about this as we run. (Runners always think about something, to distract themselves from the pain of running.)

* * *

THE RACE BEGINS near the Museum School and the museum itself, winding around the Fenway and along the river to the Public Garden—following Frederick Law Olmsted's "Emerald Necklace" of parks around Boston, in a race for charity. My favorite stretch is the run down Storrow Drive, usually seen from inside a machine in thick traffic but now blocked off just for us. I run close to a thousand miles a year, in a long experiment from which I must now report the unhappy findings:

Slow and steady never wins the race.

And just ahead of us in the crowd, by chance, is the mayor and a few of his more athletic appointees. Running, for me and I expect for the Mayor, is a reliever of stress. Sometimes the spectators help: as we pass a large group along the Fenway, one calls out, "Hey, Mayor, watch out for the potholes!"

Looking down at the road and the potholes passing beneath our feet, I speculate on how long one spends at each point. The answer is no time; a point in geometry occupies infinitesimal space, and we pass it in infinitesimal time; so beneath our feet there is no present, only a "becoming." Alfred North Whitehead's "Principle of Process" says:

> That how an actual entity becomes constitutes what that actual entity is; so that the two descriptions of an actual entity are not independent. Its "being" is constituted by its "becoming." [2]

So are we all, becoming. And ahead, each farther step anticipated becomes progressively less certain, and one cannot see at all around the next bend. Even the very next step is not completely certain (given the potholes), but new information there awaits discovery. The runners by my side share in the discovery, enhance its interpretation, help me test our old hypotheses against new evidence. This is what Whitehead says:

> ...the true method of philosophical construction [is]
> to frame a scheme of ideas, the best that one can,
> and unflinchingly to explore the interpretation of
> experience in terms of that scheme....
>
> Rationalism never shakes off its status of an
> experimental adventure. [3]

Whitehead was a mathematician and knew that no process can be entirely arrested, that one cannot stop a particle or an occasion to examine it completely. Points in time, as well as in space, must be considered infinitesimal and cannot therefore be pinned down. One can define a point only as arbitrarily small and then go on. My view of the pavement (despite my slow pace) is never fixed, my position keeps changing, my stopwatch keeps recording new "presents."

Our most cherished theories, our laws and most confident findings, having passed every test we can devise, yet await other trials as we round new bends. In this sense the hypotheses in these letters must be taken: not as tentative postulates for the sake of argument, not without confidence, but as the best we can see from this stretch of road, offered as worthy of trust and commitment but always open to new tests. Like Harrison's "masks of the Universe," each new set of insights seems to satisfy the evidence at hand but does not exhaust discovery. [4]

Well, perhaps we shall find the real world just around the next bend. Or perhaps it lies behind us, and we missed it. At the half-way point we turn the corner of Charles and Beacon streets, and I look back on the miles I have run.

* * *

WE HAVE PASSED the Arlington Street Church where I sang for a summer, and I can see the corner of West Cedar Street where Abigail Eliot grew up as the twentieth century dawned, and beyond it the State House and our old TDC office on Beacon Hill.

Stretching back over the miles are other turns that have made, as Robert Frost said, "all the difference": along the English hedgerows above Gloucester Cathedral, around the Musée Marmottan in Paris, across the Southwest desert to Hawikkuh and Hard Scrabble Wash, out the Harvard Observatory path from the TCC laboratory, into Middletown Prison and Newfern School and the Metropolitan State Hospital. The road runs along rough streets in Roxbury and the South Bronx, up the elevators in office towers, down the long Pentagon corridors, through the Bolton cemetery, over the Green Mountains, into the Fitness Club, around Monhegan Island and Tilden Pond, across the MIT campus, and into Emerson Hospital where it started.

These places, experienced momentarily in their becoming, constitute my "real world": events in which one may witness the greatest strivings of the human spirit. Running by my side have been the doctors and patients, guards and inmates, researchers and students, ministers and parishioners, coaches and performers, friends and family who populate this real world and call us to experience it.

At each turn we also encountered the art or writing or building of those who had passed this way before—those who experienced the striving and those who understood it, who could see beyond it. The poet and artist, scientist and philosopher, designer and builder, teacher and healer, illumine each place with their great, clear seeing— at once universal and special in these places, attuned and resonant in the presence of these deep strivings. Here they share with us that coupling of discovery and hypothesis that supports our becoming.

Like the runners at my side today, we all pass through our real worlds at a pace that supports not arriving but becoming. One learns from the residents we pass and the visitors with us, from the scientists and artists and builders who preceded us, from those who hope mightily and those with no hope, from the imprisoned (in many ways) and the free (in some ways.) One learns from inspired insights and from missed diagnoses and from confused frustrations. Each reader of our shared experience has presented a transparency to realities we would have missed without him or her, and we stand in wonder at what some of them could see. Now they have raised the place on which we can stand, given us a farther horizon and better instruments, and we have the chance to do the same for those who follow us past the museum and around the Emerald Necklace and down the highway we share.

* * *

So WE GO on learning more, testing and refining each hypothesis, building together upon the foundation we share, becoming. The road behind continuously lengthens, building one's stock of "used information" and, to the extent the road is shared, building context for further discovery. Each step is taken with hypotheses in mind, to be tested on the road ahead. We discard some with our growing perception, with our shared testing, and we keep some, modifying and retesting and formulating new hypotheses. This is the process of all science and all philosophy.

As we look back we can identify errors and discarded notions among even the most respected thinkers, but they lighted our path and helped us choose our turns. So we too shall be found wrong and irrelevant in many of our most cherished notions—whether we consider them scientific or philosophical—and many of our hypotheses will be discarded. Yet we need to share them and to commend their critical review to those who join the race farther down the road.

Many of our fellow runners hesitate, in great fear that advancing knowledge threatens what we hold sacred, that solving mysteries reduces the realm of beauty, that faith and love cannot be rational, that our sense of the holy may be obsolescent in a world of technology. I find no such erosion along this road; in fact the opposite is evident from these encounters.

The science that enshrines a commitment to truth reveals the deepest beauties, finding order in places that were unknown or thought to be chaotic. Many who have devoted their lives to science testify to their sense of awe at the great beauty that lies at the heart of nature. They trust, moreover, in an intrinsic "elegance" of nature, and this faith has been rewarded again and again in advancing understanding. The experiences of order, of symmetry, of finding deep and hidden relationships, of consistent metaphors and analogies—all are deeply scientific, as well as beautiful. The trust that we can understand is a trust that we can grow. The commitment to truth enables reliable communication, supporting mutual benefit and the experience we call love.

* * *

So I AM confident that the hypotheses considered here are consistent with our great religions' central teachings and with the advancing scientific understanding of our universe and with the art and literature that has refreshed our pace and lighted our way and with the deepest needs and highest aspirations of those encountered in their striving-places, along this long road.

* * *

WE LOOK AHEAD and see no end to our road, and we look back and see no beginning. We look to each other and find a becoming.

Notes

[1] See the following letter, The Board Room, for an extension of this concept of time, and the letter from Radio Communications Corporation for the underlying theory of information.

[2] Alfred North Whitehead, *Process and Reality*, corrected edition, ed. by David Ray Griffin and Donald W. Sherburne (New York: Free Press/Macmillan, 1978), 23.

[3] Ibid., *xiv* and 9.

[4] Edward Harrison, *Masks of the Universe* (New York: Macmillan, 1985.)

The Board Room

T HIS MORNING I am being gobbled up by what Bob Mueller
calls the "heavy breathers" who represent our major inves-
tors. [1] We are in a magnificent board room, high in an
office tower overlooking the bay, but the view and the food have lost
their taste for me. I have served on many boards, but never developed the political skill of maintaining control over opposing interests in the face of sustained bad news. By the end of the meeting I
will be deposed as president, and the company I started will be in
the hands of a management committee of heavy breathers.

This leads me to reflect (after a suitable recovery period) on the
nature of influence. All the board rooms in which I have met over the
years, from the ornate to the simple, controlling large private investments or modest grant support, have been populated by "influential"
persons. Our proceedings, moreover, projected influences far beyond
the chairs in which my colleagues sat. I can still walk into many of
those rooms and look at the empty chairs, remembering the impact
that their occupants had on me and others. Lenders and borrowers,
physicians and patients, trustees and managers—now their chairs are
empty, and I search them for signs of the vitality that I remember.

Perhaps we can find the source of influence in the elemental
particles that make us up. I have been reading Leon Lederman's
amusing and instructive account of the search for the ultimate elemental particle (the "God Particle," he calls it [2]), and I am struck
by the fact that, in the micro world, the particle *is* its influence.
Lederman helps us get our minds around the concept of a particle

that has great influence but no size—a point particle, with charge
and mass and spin (its quantum characteristics) but occupying no
physical place. He says he thinks of Lewis Carroll's Cheshire Cat.
The cat gradually shrank down as Alice watched until nothing was
left but its smile. (Carroll himself was a mathematician, and his alle-
gory had substance as well as bite.) Thus, says Lederman:

> ...the electron is real. Probably a point particle, but
> with all other properties intact. Mass, yes. Charge,
> yes. Spin, yes. Radius, no.*
>
> Think of Lewis Carroll's Cheshire Cat. Slowly
> the Cheshire Cat disappears until all that's left is
> its smile. No cat, just smile. Imagine the radius of
> a spinning glob of charge slowly shrinking until it
> disappears, leaving intact its spin, charge, mass, and
> smile. [3]

George Johnson, who writes about science for The New York Times,
considers particles like the electron to be "all information":

> A marble has mass, color, size. But an electron is
> mass, spin and charge. A particle is completely
> defined by its quantum numbers. It is all information.
> Spin ½ plus 1 unit of negative charge (1.6021892
> x 10^{-19} coulombs) plus a mass of 9.1 x 10^{-28} grams
> is an electron. These are not just labels or qualities
> exhibited by something underneath. There is nothing
> underneath. [4]

We are not likely to find direct determinants of human personality
among the random mutations and other intrusions of the micro
world into our experience, even though we are made of Lederman's particles. We do share, however, an instructive aspect of our

---

* But see footnote, p. 73

"influence," one that appeals to me especially as I regard the empty chairs in our board rooms.

Look at the missing occupant, they seem to say, as you might look at an elemental particle, an electron perhaps, or a photon. What you see is the influence but not the location. Look at the person who sat here, not as a constrained object in physical space, one we can measure and define and thus limit, but as an expression of real qualities. [5] In the micro-world of elemental particles, the real qualities are charge, mass, "spin," and other quantum numbers. In our experience in these meeting rooms, they are commitment and transparency and relation. Like the electron's influence and the Cheshire Cat's smile, these are the qualities that remain when we give up the attempt to capture and hold an individual in her chair, i.e., in a defined space or time.

* * *

PAGELS'S QUOTATION of Einstein, mentioned in the letter from the Metropolitan State Hospital, bears repeating:

> A human being is part of the whole, called by us "Universe": a part limited in time and space. He experiences himself, his thoughts and feelings as something separated from the rest—a kind of optical delusion of his consciousness. This delusion is a kind of prison for us, restricting us to our personal desires and to affection for a few persons nearest us. Our task must be to free ourselves from this prison by widening the whole circle of compassion to embrace all living creatures and the whole nature of its beauty. Nobody is able to achieve this completely but the striving for such achievement is, in itself, a part of the liberation and a foundation for inner security. [6]

Heisenberg's uncertainty principle says that we cannot pin down both the position and the momentum of anything with arbitrary accuracy. If we hold it still to define its position, then we cannot say where it is going: as soon as we make a location measurement, the domain of its future positions spreads out wildly.* If instead we track the particle's momentum (a "vector" quantity involving direction as well as speed), then we cannot precisely define its instantaneous position.

Just so, I believe Einstein and Heisenberg would agree, we cannot pin down a person's future course and present limits, but must yield to the freedom of intelligence, as we did to the uncertainty in a particle's momentum and position. The person's "influence" may be considered as her freedom to reduce uncertainty, to affect in some degree the probability of future events. (This is one of our evidences of life itself.) None of the board members, however great her resources or credentials, can influence the past. Nor does her influence touch the (infinitesimal) present. It is only the likelihood of future events, as they play out, that bears the mark of influence. Perhaps this seems obvious, but our experience of time is still a puzzle.

Consider an arbitrarily narrow slice of time called the "present"—to make it understandable, let's say a millisecond. The amount of information any of us (or all of us together) receives in a millisecond is actually quite large, but it is finite. (TV channels, for example, have bandwidths in the megahertz range, so we can resolve thousands of phase and amplitude changes in a millisecond.) As we narrow down the slice from a millisecond to a microsecond and then a nanosecond, the amount of information we can process keeps

---

* This can be demonstrated dramatically by gradually narrowing a slit through which we pass electrons or photons, even one at a time: at first they impinge on a small target beyond the slit, but as it narrows (as we determine their position more and more closely) they spread out to a wider and wider range of destinations. Although we can predict their cumulative spread, we cannot tell where any one will go.

shrinking. Eventually it becomes arbitrarily small, as the amount of time allocated to it becomes arbitrarily short. We must define the "present," I think, as that infinitesimal point at which the smallest quantum of information may be transformed from mystery to memory.

The arrow of time is that direction in which information passes from unknown to received to remembered. I will call these phases of information "latent," "actual," and "remembered." Shannon's formula measures only the first, because the receipt of information eliminates its uncertainty. (Intuitively one wants to associate information with increasing *certainty*, and there is some confusion in the literature on this point. The true measure of information in any message, however, must be the amount of *un*certainty it contains. [7]) Once actualized by intelligence, this information passes from mystery into memory. Our "present" experience continuously moves into the realm of latent information, adding to memory as a wake in the infinite sea of information:

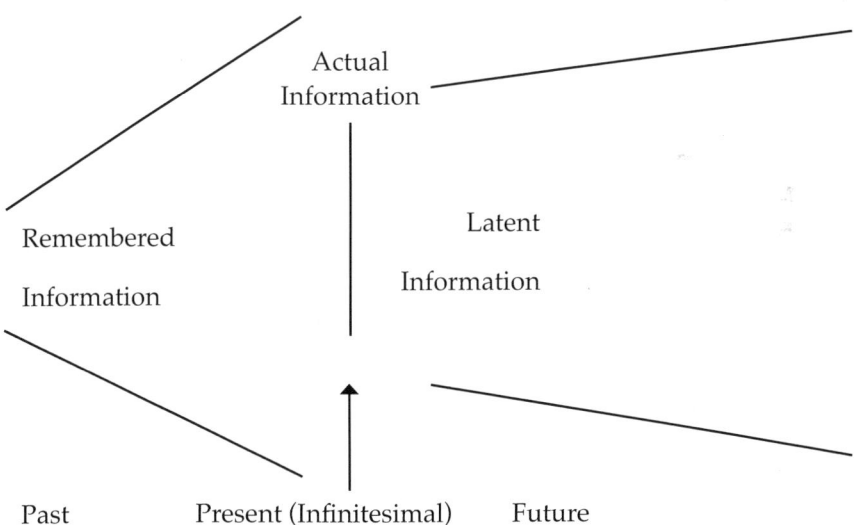

Actual
Information

Remembered

Information

Latent

Information

Past   Present (Infinitesimal)   Future

(I have drawn this as a two-dimensional time plane, rather than a one-dimensional time line, to suggest the amount of information as a vertical axis.)

* * *

This is both the psychological and the thermodynamic explanation of time. Psychological because all experience is perception, thus processing of information. Our psychological "direction" of time is given by our receipt of new information and its conversion into memory. Our psychological "speed" of time is inversely related to the amount of new information we process. (Time seems to move faster as we age or as we retrace a familiar route, slower when we are young and encountering unfamiliar territory.) The thermodynamic direction of "future" is toward equilibrium, which is the most probable state because it has the greatest number of possible ways to get there.* The thermodynamic number of possibilities increases in the same direction as the psychological and informational arrow of time. Thus "latent information" *increases* as we proceed toward equilibrium, whether actualized by intelligence of not. And intelligence does not exhaust the store of latent information because information grows along the same "time" line.

The Second Law of Thermodynamics says that entropy must increase with every transaction (except those few that are entirely reversible): we generate more disorder (usually in the form of heat) than we dispel. Every bit of information "actualized" comes at the expense of more than one bit of added uncertainty in the environment, because we have to expend energy to receive it.

My picture is two-dimensional, however, which suggests we might consider an orthogonal component of time—a third dimension, rising "out of" the page toward the reader or sinking down from the page away from us. In such an extension, "toward" could be considered the orthogonal past and "away" the orthogonal future. The present, being infinitesimal, exists in all dimensions; it is like

---

* Thermodynamic entropy is proportional to the log of the number of microstates, always maximum at equilibrium.

the pin of a hinge, around which one could rotate the time-plane to any angle. In our experience, however, we must move along the same plane as our past; thus the two-dimensional diagram suffices for customary experience—we are constrained to the plane of the page. If we had no past, however, i.e., no memory, we would have more degrees of freedom, starting from our present-line and moving along the page or at any angle to it. This may be the case for the quantum particle in a superposition of states: until "measured," it has no past (no memory) and therefore its time-arrow is undetermined. (In fact, some representations of the possible histories of quantum particles include paths going "backward" in time, as can be seen in some Feynman diagrams.)

Although you and I do not experience such orthogonal time (just as we do not experience more than three dimensions of space), we can easily express it mathematically and it works in all our normal time-based formulas. Therefore a different world-line, a whole different experience, is theoretically possible, following an orthogonal time-line. One can find "imaginary" time (which I consider equivalent to orthogonal time) already in the literature, because it is mathematically elegant and convenient in explaining certain quantum events and the strange possibility of instantaneous quantum communication. [8] We should see along this same perpendicular dimension what I will call "orthogonal information." (Actually any angle to our time-plane is mathematically allowable, but only the ninety-degree rotation has no "shadow" on our own experience, having a cosine of zero.)

Is this just a game, a trick, or might the concept of orthogonal information have an important reality? There are too many cases in scientific history of mathematical curiosities predicting physical realities for us to dismiss this one. If we knew all, there would be no time (in any dimension), since without uncertainty there is no flow of information. Thus we may consider truth, as containing all information, to be timeless and to occupy all the space in all rotations of the "present hinge."

I do not know what developments you and your children will witness in this direction, but I think their pursuit may influence future understanding of our place in the universe and our connections with life generally.

* * *

So my fellow board members, your influences remain with me long past adjournment. We do have influence over what we call the "future"; i.e., we can alter the probabilities of events in the field of latent information. We do not have the ability to pin down limited definitions or roles for each other. We draw the "past"—remembered information—behind us along our plane of experience, as our present line moves into a sea of infinite possibility. Perhaps we shall find a way to turn on this hinge called the present and move along a new plane, with the freedom of quantum particles devoid of memory. Meanwhile the great improbability of life that we already express, evolved along our unique time-plane, affords us the capacity to influence what lies ahead.

Now the chairs are empty, including mine, but they were never the seat of our enterprises. It was our collaboration that set a future course, our sharing of commitments that vitalized our endeavors. This is the legacy of the communication sustained among us; and I believe it is neither lost in our apparent separation nor threatened by our separate pains.

Notes

[1] Robert Kirk Mueller, *Behind the Boardroom Door* (New York: Crown Publishing Group, 1984.) Bob served on this board with me and has been a steady source of good advice and good humor. He has written a dozen or so books on boards and corporate governance and was chairman of Arthur D. Little when I first met him.

[2] Leon Lederman and Dick Teresi, *The God Particle: If the Universe Is the Answer, What is the Question?* (New York: Bantam Doubleday Dell, 1993.)

[3] Ibid., 142.

[4] George Johnson, *Fire in the Mind: Science, Faith and the Search for Order* (New York: Vintage Books, 1996.)

[5] See the letter from the Marmottan Museum, and the lessons of Monet and Gould about the indistinct boundaries of the "individual."

[6] Albert Einstein as quoted by Heinz R. Pagels, *Perfect Symmetry: The Search for the Beginning of Time* (Simon and Schuster, 1985), 362.

[7] See the letter from the Radio Communications laboratory, and its description of Shannon's information theory.

[8] The full potential of quantum communication (and its obverse, quantum encryption) is not yet known as I write, but it is at the forefront of "practical" applications of contemporary theoretical physics. See, for example, Dik Bouwmeester, Jian-Wei Pan, Klaus Mattle, Manfred Eibl, Harald Weinfurter, and Anton Zeilinger, "Experimental Quantum Teleportation," *Nature* 390 (1998): 575–579.

The South Bronx

HERE IS A neighborhood of death, and of life, the choice held up in terrible opportunity each night. We do not often look up: all around the vacant windows stare down, in mocking repudiation of our purpose here. They are too many to count, dark even at noon, echoing obscenities in their hollow souls, witnesses to the most grotesque and perverse evils acted out within and on the stage below. One cannot endure their stare and hang on to innocence.

These eyes of deserted buildings threaten us with more than present risk. They look deep into our own pasts, find our hollow places, strike jarring resonances within us. I hear their reverberation behind my own facade of principle and deed. It is an echo of my infidelities, of missed chances to help, of being untrue, of hurting others. The stare of deserted buildings summons all our ghosts of shame. I cannot hide from them here, cannot suppress my memories or go back to right them, and that emptiness rings and aches when struck so sharply by this piercing stare. For these hollow buildings confront us with the very archetype of desertion.

So I come to the South Bronx with a personal as well as a corporate agenda. I think they are related. I find the beginning of forgiveness in remorse and commitment, but the sense of walking away from the old person, of healing somehow what I have broken, this re-integration requires others. I find them here.

Our corporate agenda is to support economic development in a program called "Bronx 2000." It is a training and small-business support program, trying to get unemployed residents to believe in themselves and each other. Our hope is to demonstrate together that self-interest is other-interest, that commitment and trust in each other and in truth is a way to live, irrespective of circumstance. We bring some skills, some time and some investment—not nearly enough of any—and we take away a maturing in ourselves.

The programs of our little company take me to the South Bronx and north Philadelphia and Roxbury, Massachusetts, experimenting with self-sustaining businesses that might open a few doors for those ready to walk through them. Tonight I am teaching electronics to aspiring technicians at OIC (a branch of Opportunities Industrial-ization Centers, organized by a Rev. Leon Sullivan in Philadelphia.) I learn, of course, more than I teach, and I meet many whose com-mitment to this neighborhood and its people dwarfs mine.

The South Bronx is a good place to test our hypotheses. Distinctions are sharper here, the contrasts of good and evil more focused, and there are fewer places to hide. The nights are ferocious. Violence stalks them with and without purpose, drawing fear and pain of every kind behind. The streets can be as brutal and ignorant as a medieval slum, and there is no glossing over their dark reality. There is coming and going from the suburbs, and there are prayers and easy generalizations, but the street-level world is unmoved. Like the prisons and mental hospitals, it seems impenetrable by anything gentle. Yet we find here remarkable examples of sacrificial commit-ment, of unconditional relationships, of courageous giving, that help us understand what we mean by "good," as well as by "evil."

To be committed to truth and the benefit of others is one thing, but to trust in them—for faith requires trust as well as com-mitment—is to stand straight in a chill wind of lies and malignancy. Our faith that love is stronger than hate is easy to sing in choir, but here it takes a continual reminder on rational grounds.

* * *

ARE WE, AS Whitehead says, a "becoming" people? If this has meaning, it should be testable on logical grounds. One test would be whether evidence can be demonstrated of substantial increase in (a) the capacities we associate with living, and (b) the qualities we associate with being, with identity. Can we show a general growth of freedom? of organizing and building and bringing order from chaos? of communication? Surely yes over millennia, perhaps over most generations; thus test (a) appears to be met.*

The enhancement of life-capacities speaks, however, only to the quantitative part of becoming. It is essential, of course: those who lived through the ice ages and bore children in pain and survived a brutal world, those our ancestors bought us the freedom and comfort to contemplate, and paid a high price. We stand on their shoulders. Their gift makes us free, even possible; but it does not make us good.

The qualitative element of living is what we make of it, and in particular how our identities develop. This is test (b.) We know each other not simply by the capacities of life—we all have those, in greater or lesser degree from time to time—but by the commitments we make, the transparencies we show, the relationships we nurture. If we are a "becoming" people, we must testify to our qualitative, as well as our quantitative growth.

The logical direction of becoming would be the development of a fuller expression of life and identity. Our means has been clearly demonstrated over these millennia: it is to collaborate, to choose relation. That is the link between quantitative and qualitative living. We are given the capacity to explore together, to build together; we are given the freedom to choose, and an ideal means of being together: by communication.

---

* This conclusion must always be qualified by the recognition that the conditions and capacities of "living" in much of the world regress frequently.

What is not given—how we use these capacities—is the qualitative side of "becoming," the identities we express. A becoming people climbs together the ladder of communication. The logical direction of becoming is a joining of capacities, a sharing of ideas, a mutual helping, a freeing of each other from the old pains and limits, a making of art and science. This allows us to grow both quantitatively and qualitatively, to become a stronger and a better people; but we seize the opportunity infrequently, as the eviscerated buildings of the South Bronx eloquently attest.

* * *

HERE ON THE streets there is more evidence of tearing down than of building up. Freedom has tight boundaries on the streets, and communication does not often run deep. Yet these capacities can be found in all, from time to time. And so can commitment, and transparency, and relation, sometimes in deeper and clearer ways than in more comfortable circumstances. Both living and being, both our quantitative and our qualitative manifestations fluctuate. We do not meet any pure saints at home or any unredeemable sinners on these streets, nor do we find any who own life without discontinuity. So we have not become. But are we becoming?

In these neighborhoods the logical direction of becoming is starkly seen. It is not a sophisticated argument or a gentle philosophy in the South Bronx. Here we can meet women and men whose commitment shines in bright contrast to the gray rubble about them, whose relation to their patients and clients and little brothers and sisters requires no great perspicacity to discern. Here is the opportunity for transparency, where needs are great and the fog among us is thick from fear and want. Here are my heroes, men and women, counting more embraces than abuses, working through the night while most of us sleep. Their self-interest is truly in other-interest.

What so clouds our seeing of self-interest in others? Are these heroes of mine immune from our myopia, somehow superior in vision or courage? The streets are my classroom, and gradually I learn to distinguish behavior from character. Those with little hope experience short horizons, within which long-term benefits are remote from immediate needs and cooperative action has a more remote reward than personal gratification. In our comparative comfort we wonder at the fatalistic, self-destructive behavior encountered here, but it takes some faith for any of us to see the long and the collective view. That faith—that trust and commitment, that extension to others—comes much more easily to us when we have reason to hope, to expect a better tomorrow.

There is, however, a rough neighborhood in each of us. As we do not express intelligent life continuously, we do not express our best identities consistently. Each of us can hear within ourselves an echo of the screams that tear the streets. And each of us—resident and visitor—has an equal opportunity to improve our behavior each morning.

Clearly our quantitative progress has raced ahead of our qualitative. These concrete shells are home to many creatures but little love. The ability to discern order and to choose it supports not just building but our experience of beauty and love. Are we using it better? Our ability to heal, to educate, to dispel mystery, to ease pain, to communicate, have advanced, but our commitments are still shallow, our relationships no stronger than those of medieval times.

The progress of any people accelerates when they can share what they learn, build together, communicate truthfully, show each other lasting realities, commit and trust and relate. Our eradication of disease, our exploration of the universe, our building of comfort and challenge for our children, all that advances civilization, depends on mutual trust, mutual commitment, and the extension of individuality. The reality of hatred and the ability to destroy obscure progress and potential but do not change how they are realized.

This is the burden of freedom taught so starkly on the streets—that we can choose to build or destroy, to associate or isolate, to benefit or to harm. It is logical and rational to choose the former because one's self-interest is thus advanced by such other-interest. The Whole Each Other by definition includes oneself.

* * *

WHEN WE NAMED Justice Research Institute (a company originally formed by TDC employees), we had in mind some improvements to the prevailing criminal justice system and did not stop to contemplate what justice means to different people, what elements of justice might be wise to strengthen and how. Our general concept and practice of justice has religious roots in much of the world, growing out of some combination of creed and covenant. But there is a difference, the latter being more consonant with rational faith. A covenant is an agreement to walk together, a mutual promise. A creed is something else, and may be irrational and exclusive.

Those who took up JRI's cause and built it beyond our vision understood (or learned from their clients) some finer distinctions. They teach us that "good" relates to behavior that is truthful and beneficial to others. Goodness is not derived from (or necessarily dependent on) obedience to particular rules. The evil that we see in abundance here is behavior that is untruthful or harmful to others. That is the distinction between good and evil, and our reaction to evidence of either needs to be faithful to the same distinction. Do we punish for correction and deterrence, or for revenge? For the good of all or for the satisfaction of our emotions or creeds? Do we reward to encourage better behavior or to purchase loyalty and protect our personal investments? Such questions must be confronted daily by those working in our systems of criminal and civil justice.

Each of us can strive to be committed, related, and transparent to truth and to the benefit of conscious life, to advance our collective capacity and freedom to organize and to communicate. As these goals are logically derived from a rational view of the life we share, our sense of justice should consider one's faithfulness to them. It is convenient to consider some people bad and others good, but it is not accurate. In our prison visits we often find compassion, and in our churches often selfishness. They are both pervasive and none of us is immune. That realization helps us deal with the freedom that life grants us. We need to distinguish between bad people and bad behavior. Enormous capacity dwells in each of us for good or evil, and it is just below the surface. One need look no farther than one's occasional fantasies, or one's actions when angry or frightened, or the reactions of ordinary people in an excited crowd.

The qualities that make us persons—commitment, transparency, relation—are often stronger on the streets than in the places of power and worship. We find a moral compass as steady in those of modest gift as in the much-gifted. We see commitment and trust and extension—which we call love, or faith—clearly here, and relation close and dependable. Shall we find such qualities as surely in our lab? In the Pentagon? At MIT?

* * *

I RETURN HOME with entirely new perspectives on good and evil, not just in others but in myself, and with a sense of healing. This has derived, as in all the places from which I write to you, from those encountered on the field of striving. They have been present in immediate person and in past writings, in seeking help and giving it, in learning and teaching, their roles exchanging and fluctuating. Being there to listen is my great good fortune and the authorship of these letters.

We are all drawn to the streets by mixed motives, by guilt or ego, by a hollowness echoed in the vacant buildings, by our search for meaning or comfort, for worth or forgiveness; but we encounter instead a mocking evil and impossible examples of good. In despair of conquering the evil or equaling the good, we gradually discern and accept their varying presence in ourselves. The freedom to choose between them, the capacity to be faithful to each other in a universe that sustains life through our nights, is not earned but given, by the prevailing of truth in every event. This is the grace of a rational faith.

We yearn to be somehow forgiven. Our past, however, is made of remembered information and cannot be erased by a ceremony of contrition. It is more we need, beyond remorse, beyond confession, beyond the lifting of our shame from these hollow shells into the light, it is more than this necessary but still not sufficient start of healing.

We look at each other on the streets of stark contrasts and ask what is missing, and in our coalescing transparency a way of reintegration gradually appears. One's unfaithfulness to truth and to each other suggests its own remedy: each injury was, like the abandonment of buildings in the city, a desertion, a forsaking, a severance. This goes beyond the immediate instance; unfaithfulness to one is unfaithfulness to all—both the offender and the aggrieved extend in influence and relation. This makes the offense worse, but it makes a remedy possible. An opportunity to reintegrate, to heal, to connect is presented in each future encounter with others. We find pardon in faithfulness to that which we really deserted, to truth and to the whole each other.

* * *

IN FACT WE cannot be whole alone. It is when one enters solo on this stage in robes of righteousness that one is most mocked by the forsaken structures all about. The pride of doing good that one carries into these shadows becomes a great burden, a delusion, an encapsulation, a slick film on our grounding. Only in opening to a wider self, only in transparency and relation can our commitment find purchase on these slippery streets. In this one need not be a Mother Theresa to do good. The judgment of truth concerns the faithfulness of application, not the achievement of result. Every encounter, in the most routine of daily professions and activities, is an opportunity and a challenge. One may reach more and farther by a consistent transparency than by an impossible example.

* * *

So WE CAN lift our eyes and look with confidence upon the vacant windows in the South Bronx, at the empty hooded visage of death in our cities, and in ourselves. Once we confront the capacity for evil in each of us, we can trust the good in all of us.

Northfield Mountain

THERE BREATHES WITHIN this soft green hill a giant troll, grumbling away in all seasons, a copper-iron creature drinking the Connecticut River nearly dry at night and spitting it out again each morning. He strives all night against the mightiest monster in our universe, and loses every night, and exhales his long slow breath all day in rest, then gathers himself to fight again come evening. His adversary is ours, and we have named her Entropy, or sometimes Nature.

She dissembles here as friction, there as erosion, here decay and there collapse, and everywhere heat and equilibrium; but she is Nature and she is Death. We have set our troll, like Sisyphus, to an endless task upon this hill for a narrow purpose that I understand well enough; but the servant reminds his master that we strive alike against a common foe, against inexorable disintegration. So I come to Northfield Mountain with two eyes, and see two stories acted out.

* * *

THE FIRST STORY is easy to tell. Northeast Utilities has dug an enormous cavern in this hill above the Connecticut River, and a large reservoir at its top. At night, when the public demand for electricity is relatively low, we turn on four big electric pumps installed within the cavern, pumping over 12,000 acre-feet of river water up to the reservoir; and when demand peaks during the day we let the water pour back down through our pumps, reversing their spin to

237

generate fresh electricity. This gives us almost 8,500 megawatt-hours of electrical energy when it is most needed.

Our problem is that we get back each day from the generators only some 100 kWh of electricity for every 135 kWh expended using them as pump motors at night; the remainder is dissipated in the heat of friction, electrical resistance, irregular flow, and other random processes that we cannot stop or control. This heat of dissipation (divided by the temperature at which it was generated or transferred) is the most familiar form of what we call "entropy." So tonight we shall spend more energy pumping the water up than we can recover tomorrow letting it pour down again. We perform this Sisyphean task simply because it reduces the total generating capacity needed by the New England Power Pool to meet peak (daytime) demand; it saves money even as it loses useful energy (which goes into entropy.) The same is true of other forms of energy storage. [1]

I am among the least of the officers listed in Northeast Utilities' annual report, inheriting a vice-president's title when my company was sold to NU some years ago. Our management of energy storage and efficiency has become the basis of a whole industry, whose members are called energy service companies or ESCOs. With exquisite timing, I had founded our ESCO at precisely the peak of world energy prices in 1982, whereupon its value plummeted faster than Northfield's river water. My reward has therefore been more honorary than financial. In any event the industry prospers now, and does generally useful things for the environment and the economy. It also offers an important philosophical lesson on the way our universe works, and that is the second story told by the Troll of Northfield Mountain.

* * *

WHEN WE WORK on improving energy efficiency or energy storage, as the ESCO industry does, we strive against entropy: the inexorable tendency toward equilibrium, toward disintegration of all structure, toward deterioration of directed energy (work) into chaotic energy (heat.) There is nothing magic or even mysterious about this "second law" of thermodynamics; it is simply a natural trend toward the most probable state of all things. Structure is improbable, chaos probable. Sharp distinctions and steep gradients are improbable, equilibrium probable. Knowledge is improbable, ignorance probable. In each case we have to expend far more energy to sustain the former than to allow the latter. There are simply more ways to get to a disorganized condition than to an organized one, so its probability is higher—very much higher, as it turns out.

Whenever any system or mixture or gradient "settles down"—reaches equilibrium—we find that distinctions have blurred, differences smoothed out. A glass into which yellow and blue fluids are poured from opposite sides quickly turns green, and stays that way, because there are enormously more "mixed up" arrangements of molecules than the few that will still look yellow and blue. All the molecules keep moving around and bumping into each other, under the influence of heat-energy; the probability that the yellows and the blues will ever again be found arrayed on opposite sides is extremely low. (In just a thimble of a simple gas at ordinary temperature and pressure, the number of such molecular collisions is on the order of 1,000,000,000,000,000,000,000,000,000,000,000 every second, which gives some idea of the low probabilities associated with maintaining distinct, segregated arrangements.) We could force the yellow and blue molecules back to their original places with a sufficient mechanism for detecting and catching and moving them, but it would require hugely more directed energy than the random heat energy that keeps them mixed up.

Thus it takes more energy to maintain or restore an intended state than to let it fall to its lowest level: when a teacup shatters or a blossom falls, we cannot put it back together without enormous effort. When a clock's pendulum swings, it cannot regain its initial height without our winding a spring to re-inject the energy it lost to frictional heat. When I run downhill, I never recover all the energy that I expended climbing the hill. The Second Law says that all clocks will eventually stop, all blossoms decay, all runners tire, all structure break down to its lowest state. There are no "perpetual motion machines," nothing runs forever, there is something lost in every transaction. We can reduce this entropy locally, by reassembling pieces or organizing processes or extracting heat; but the reduction of local entropy is always accompanied by a greater increase of entropy in the surroundings. Nearly all the processes we see are irreversible without substantial addition of directed energy.

The same phenomenon operates at the level of quantum particles, and gives us the statistical definition of entropy: it is the natural logarithm of the number of different microscopic arrangements in which a system could be found (multiplied by a well-known constant.) [2] This number, called the thermodynamic probability, is huge for systems of everyday sizes. Moreover, the number of possible arrangements that could produce the "most probable state" of any system at equilibrium is nearly indistinguishable (by most of our instruments) from the total number of all possible arrangements—other states are possible but have vanishingly small probabilities. (The "system" most often considered is an isolated container of a gas or liquid of defined characteristics under defined conditions; but the concept is expandable to include larger objects, machines, people, planets, and universes.)

Again it is just a matter of chance: the most likely state is always that which has the most ways to get there—unless we inject a lot of focused, organized, directed, purposeful, coherent energy to choose and sustain a different state. This is how the universe works.

* * *

So THIS IS the nature of things, this inexorable dissipation of order. We and our Northfield troll, and all of us, strive daily against the great monster of entropy, which continuously breaks down our organization and coherence. We can win local battles but she prevails in the global war. It is a gloomy forecast. If our universe continues to expand forever (considered likely at the moment), its death will be ultimate disintegration, the decay of all directed energy and all structure. (Of course if we're wrong and we live in a "closed" universe of sufficient mass/energy density, the forecast is not notably cheerier, involving a "big crunch" of everything back into a gravity-induced fireball.)

Who or what, then, strives most mightily against entropy's consumption of order? It is not our troll or other machines that give her greatest affront, nor structures nor gradients nor high temperatures, all of which she conquers (given sufficient time to reach equilibrium.) It is we. It is intelligent life. This conclusion leaps from the evidences of life as we have considered them:

- The capacity to organize, to build, to structure, synthesize, cooperate, integrate
- The freedom to choose, to alter probabilities
- The ability to communicate, to exchange information; i.e., to reduce uncertainty*

This is the very antithesis of entropy, the Great Improbability, the most unlikely thing in the universe, the "far from equilibrium" condition celebrated by Prigogine and Stengers. [3]

---

* In fact, one may derive the formula for thermodynamic entropy directly from information theory. If we had free access to all information (including that at the quantum level), we could recapture the heat of dissipation, sustaining reversible engines and putting fragments back together without losses. It is not mechanics but ignorance that ordains the Second Law.

That such an improbable phenomenon should evolve against the imperative of entropy has, of course, occupied the contemplation and imagination of great and not-so-great thinkers in all ages, including ours. Many find in such an evolution a necessary Design, a Creator, or a strong "anthropic principle" at work. [4] Others point to the vast time scale of evolution and the possibility of many outcomes, seeing the accident of human evolution as entirely consistent with probabilistic events and natural selection.

I think the argument is enlightened by raising its plane to the level of intelligence generally, above the mire of earthly evolution and the notion of exclusively anthropomorphic intelligence. Where does intelligence itself come from, and is it locally evolved and particular, or does it exist of its own self? Can we consider intelligence to be a "Platonic" reality, primordial and intrinsic, not invented by ourselves? Roger Penrose speaks of having "direct contact" with Platonic concepts like deep mathematical ideas, and suggests that Mozart experienced a similar contact with musical concepts:

> I imagine that whenever the mind perceives a mathematical idea, it makes contact with Plato's world of mathematical concepts... one's consciousness breaks through into this world of ideas, and makes direct contact with it... When mathematicians communicate, this is made possible by each one having a *direct route to truth*, the consciousness of each being in a position to perceive mathematical truths directly, through this process of "seeing"... communication is possible because each is directly in contact with the *same* externally existing Platonic world! [italics and punctuation Penrose's.] [5]

Most of us are Platonists to some extent. A majority—scientists and others—would take Penrose's side, I think, in accepting an intrinsic reality to some set of "deep truths," which are there whether we find them or not. [6] Fundamental mathematical concepts might receive the most votes for inclusion in such a set. The base of natural logarithms "e," for example, is certainly non-trivial and might not be discovered by all civilizations, but it's clearly "there." When found, we have the sense that it was waiting for us; we did not make it up, and it does not go away if we ignore it. It seems hard-wired into the basic structure of the universe, or at least our universe, and we expect its properties and utility would be invariant in time and place.

What else qualifies? The more aggressive Platonist would include the concept of a wheel, a chair, perhaps an Irish stew, in her not-so-exclusive set. Most of us would probably reserve membership for those profound concepts that we feel are independent of a particular world's evolution, that are true in every event and important in how the world works, that have a reality of their own; concepts that are discovered, not invented. I would sponsor for membership, for example, concepts like symmetry, transparency, order, perhaps entropy and relativity—but how about important philosophical or artistic concepts, for example information or beauty?

Although some would maintain that information and beauty exist without our perceiving them, my sense of Platonic reality is that both depend on intelligence. Information, as the reduction of uncertainty, exists only when "actualized" by intelligence, when exchanged. I would say the same for "beauty," that it exists only when perceived by intelligence. (It is said that Michelangelo believed a finished sculpture existed in the stone, waiting for him to free it. Shall we include the statue of David, then, or perhaps Beethoven's Ninth Symphony, among our Platonic realities?) How about music itself, or all of art? I believe all these concepts live in the realm of "latent information"—the infinite possibilities that are actualized only as they pass through the infinitesimal screen of the present and are perceived by intelligence, thus converted to memory. Only Truth itself, which is timeless, embraces all such potential beauties.

One would like to think of intelligence itself as an intrinsic, Platonic reality, rather than an accident of biological evolution; that it should exist independently of our invention. On the other hand, it is difficult to make an unassailable case that intelligence would pass the Penrose test of being "discovered, not invented." A *primordial* intelligence would qualify, but *our* intelligence clearly evolved. Thus our argument would have to fall back on human intelligence (and animal, and intelligent systems we invent) being particular expressions of the "real thing," like the shadows in Plato's cave. Can we find rational support for such an argument?

One may adopt either of two hypotheses—(a) that evolution alone carries us toward increasing capabilities to organize and communicate, and that we continue the progress through our sciences, or (b) that intelligence is in itself a Platonic reality, precedent to our evolution and awaiting our growing participation. Both hypotheses (a) and (b) support the finding of "meaning" in life; i.e., commitment to truth and to the benefit of each other, to growing and becoming. Nor need they be mutually exclusive.

On this matter I want to combine Penrose's instincts with Whitehead's [7], without resorting to unsupported preconceptions. Intelligence might be considered the consequent nature of truth, always in the process of becoming—truth being the Platonic, primordial reality, whereas intelligence (life) is a capacity to seek it, to integrate toward it—just as love and beauty are capacities that "become" (or evolve.) Taking this tack, one need not defend life/intelligence as a purely primordial, conceptual reality, but also as a consequent, derivative reality (using Whitehead's terminology.) It is clearly a reality.

Looking at the evidences of intelligent life, we can really see both natures: order is Platonic, whereas organizing toward it is derivative/consequent; freedom is a Platonic concept, whereas our "free will" is evolving and variable; truth is Platonic but information and the capacity to exchange it (to communicate) are consequent. Life seems both conceptual and consequent, primordial and derivative.

If I consider "Intelligence" to be a "Central Order" (adopting Heisenberg's term) exhibiting perfect freedom and holding all information, then Intelligence itself would be a primordial, Platonic, conceptual reality, like Truth. If I consider "intelligence" to be the capacity to find order and the freedom to choose it and the ability to communicate, then intelligence is an evolving, derivative, consequent, "becoming" reality. I adopt the hypothesis that it is *both*, because that seems on long consideration to be most cogent, most confirmed by all the encounters and authors of whom I write, most responsive to the "why" questions; it is the hypothesis in which I find both a compelling agreement of others' testimony and a resonance in personal experience; it is consistently reinforced from different angles of thought; it is "elegant" in the mathematical sense of explaining much with a compact and efficient concept. And it is the concept in which the broadest rational ideas of truth and intelligent life *converge*.

* * *

IN THIS CONCEPT what we have called Truth and Intelligent Life may be considered one: intelligence in its entirety would encompass all information, so the full becoming of intelligent life approaches truth itself. Truth—that which prevails in every event, passes every test, and returns a perfect answer to every question, that which contains all information—is the perfection of intelligence: where organization achieves a "Central Order" and communication is completed by the attainment of all information, the elimination of all uncertainty. The Second Law of Thermodynamics has famously resisted all assaults, but it would be vulnerable to siege by means of infinite information. Random processes are unpredictable processes: the holding of all information would make them reversible.* The perfection of intelligence would finally overcome entropy through the holding of all information, thus surmounting the last barrier to complete freedom—except barriers that are self-imposed.

The siege of infinite information upon entropy's walls could not be sustained without coherence. To be perfected, intelligence would have to be "whole"—without either untrue or opposing elements. The least falsehood, the slightest contention, and the coherence is broken. We never achieve such harmony in our experience, just as we never achieve full information, but it exists as an ideal to be sought. We are free to seek it or not; we are free to choose connection or isolation, cooperation or opposition, truth or error. Freedom allows the choice of isolation as well as relation (and of indifference and opacity as well as commitment and transparency.) Even with the attainment of all information, intelligence would not be *whole* until relation is freely and fully chosen. The final boundaries are those of "self," and the freedom from self comes only in relation. This is the qualitative election open to each of us, even in the full presence of truth—the degree of relation, commitment and transparency that characterizes one's identity.

The order and communication that characterize "life" must be chosen, must have purpose. The freedom to choose what to build and what to communicate is the indispensable third element one would seek in deciding whether a signal of unknown origin were intelligent. Without such freedom, we have only spontaneous crystals and undecipherable noise—rich in order and information but lifeless.

---

* Of course we cannot know all, even about single events, as Heisenberg showed. As a "thought experiment," however, we might imagine the perfection of intelligence approaching ever more closely infinite and timeless truth.

So the element of life that we have called freedom paradoxically permits imperfection of a whole intelligence. We cannot be wholly free without each other, yet we are free to remain unconnected. The final element in the perfection of intelligence, of life, is not achieved without the choice of relation among participants. Thus it may be said that to reach the fullness of living requires being, that truth requires life.

Such a concept of intelligent life would be eternal because Truth is timeless (the absence of uncertainty obviates the flow of information.) I take this—including the paradox of growing information being already contained in truth—to be a faithful (rational) reading of Whitehead's consequent or derivative nature of "God" becoming the primordial or conceptual nature:

> God's conceptual nature is unchanged, by reason of its final completeness. But his derivative nature is consequent upon the creative advance of the world. [7]

The rational thinker may substitute "Truth" for "God" and find this observation entirely logical. If we subscribe to a view of Truth that is intrinsic, primordial, pervasive, and independent of our observation (i.e., Platonic), and if we are willing to associate such qualities also with the whole of Intelligence, then we have a God for the rational.

In struggling with this paradox, I would say that we have been too conditioned, too focused on the "how," the mechanics of intelligence, and thus distracted from the concept itself. We ask, How do brains work? Can computers learn to think independently? Was there life on Mars? These are important and interesting questions, but they miss the issue of what intelligence really *is*, what it can do, how it might integrate, what its nature and prospects are when abstracted from particular support systems. Surely nodes and networks of intelligence beyond ours exist. [8] Surely they organize and communicate, make commitments, experience. Is it essential to know how they are all constructed? Probably not. Is it important to try to communicate with them? Yes. The enterprise of intelligence need not be atomized. It may have infinite potential in connection.

* * *

WELL, SAYS THE busy scientist, this is all idle speculation, and although it cannot be disproved I see no evidence to compel me to waste my time chasing it. There may be, however, very good reason to consider and to seek other means of communication. It is the mediator of all forms of life, this exchange of information, at our own cellular level as well as at the level of abstract intelligence, and great advances have always accompanied improved communication.

Heisenberg and Gödel remind us that we cannot capture the information in anything. Information, not "conserved" in the physical sense, grows without limit. (In fact we cannot prevent its growth, given its equivalence to entropy as shown by Shannon.)* So if you like mystery, be of good cheer, for it will always be with us. The comprehension of all information could be held only by the "Mind of God," i.e., by Truth itself.

A final answer to this question may be unavailable to us, because we are inside the puzzle. Ultimately we cannot discern all of a world inside of which we dwell. "Provability is a weaker notion than truth," to quote Hofstadter's paraphrasing of Gödel's

---

* Heinz Pagels at one point suggests that entropy could be considered a conserved property, because the great part of its increase in our universe has already occurred – the universe's "heat death" has already happened. He does not mean, of course, to dispute the Second Law or the inexorable growth of entropy in thermodynamic or informational transactions. Although the average temperature of the universe has declined to a few degrees above absolute zero, the part of the universe in which we are vitally interested is the far-from-equilibrium part, where entropy is still very low. [9]

incompleteness theorem. [10] We have not the capacity to understand the fullness of pervasive intelligence. We can advance together,
however, toward (and by means of) continued integration. In this
we are "becoming"—whether as Whitehead's "consequent nature of
God" or as participants in continued evolution, our manifestation of
the life qualities grows with our integration. It is not irrational to
recognize a whole that is beyond our grasp; it would be irrational
to dismiss it for lack of immediate evidence, or to cling to mysteries
that can be solved.

From time to time, in discovery or beauty or love, we catch a
new glimpse and advance our comprehension. Our Northfield Mountain troll grumbles away, always dropping more than he lifts, but
we can keep raising the Great Improbability that we share, against
Nature's downward slope. We cannot comprehend the entirety of
that which includes us, but we can continually learn. The completion
of learning would be truth itself, infinite in containing all information, and such a truth always awaits us.

All this is not a limitation of self but an enhancement. Northfield Mountain shows us a way toward freedom through inclusion
in each other. It is, I believe, an answer to the injustice of Matt's
suffering at Emerson Hospital, to the random violence in the South
Bronx, to the terror of lost identity at Met State, to Tillich's "dark
night of the soul." Once we see the possibility of our being in others,
of our becoming together, of our sharing of life, of our growing
potential in collaboration, of the advances we can make toward a
fuller grasp of truth, once we perceive this possibility and commit to
it, then we can experience the freedom and wholeness of life. To me,
this is the pervasive "field" of intelligence, in which we "live, and
move, and have our being," in the words of St. Paul—the confluence
of science and religion in a rational and hopeful age.

Notes

[1] It should be noted that energy is not "lost" but only changed from one form to another; in our case from electricity into mechanical work (done by our pumps) and thereby into the "potential energy" of water stored at the top of our hill, and finally (through our turbines) back to electricity. With every change in form, however, a significant and unavoidable portion goes into the energy of heat. The total amount of energy (electrical, mechanical, potential and heat) never changes, and we can account for it quite accurately before and after every change in form. This accounting is called the "conservation" of energy, and is a fundamental physical law in our universe, extending also to the equivalence of mass and energy in more exotic circumstances. (Unfortunately the word "conservation" has taken on a much broader popular use, such that references to "conserving energy" often mean simply reducing its waste.)

[2] This constant, known as "Boltzmann's Constant," is the ratio of the universal gas constant to Avogadro's number, quantitatively equal to 1.38×10^{-23} joules per degree Kelvin per molecule. This provides the dimensions of entropy measurements (heat divided by absolute temperature), since the thermodynamic probability is a pure number.

[3] Ilya Prigogine and Isabellet Stengers, *Order Out of Chaos* (New York: Bantam Books, 1984.)

[4] For a comprehensive description of the "wildly improbable numerical accidents" that produced a universe capable of supporting human life, see Paul Davies's *The Accidental Universe* (Cambridge University Press, 1982.) For a comprehensive review of the various "anthropic principles" and an intriguing speculation of the future of intelligent life, see John Barrow

and Frank Tipler's *The Anthropic Cosmological Principle* (Oxford University Press, 1988.)

[5] Roger Penrose, *The Emperor's New Mind* (New York: Penguin Books, 1991), 428.

[6] Ibid. See Penrose, pages 112–116, for his Platonic argument and a discussion of the opposing view of "intuitionists" like L. E. J. Brouwer.

[7] Alfred North Whitehead, *Process and Reality*, corrected edition, ed. by David Ray Griffin and Donald W. Sherburne (New York: Free Press/Macmillan, 1978), Part V "Final Interpretation," 345.

[8] Barrow and Tipler [4] spend pages 510–523 reviewing the traits or characteristics (which they call "conditions") that biologists and physicists have considered sufficient to define "life." For their definition they say that "a sufficient condition for a system to be 'living' is that the system is capable of self-reproduction in some environment and the system contains information which is preserved by natural selection." A male-female pair meets their test, as does a virus. So does a von Neumann self-reproductive machine containing a constructor and an information bank with instructions. Explicitly, however, they do not rule out non-carbon-based life forms, and in their Chapter 10 they speculate on the future of "life" as information processing, in one or many universes.

[9] Heinz R. Pagels, *Perfect Symmetry: The Search for the Beginning of Time* (Simon and Schuster, 1985), 236–237.

[10] Douglas R. Hofstadter, *Gödel, Escher, Bach: An Eternal Golden Braid* (New York: Vintage Books/Random House, 1989), 15–21.

The Emergency Room

THE PAIN HAS been severe but steady. Suddenly it leaps to crisis: sweat bursts simultaneously from all pores, the field of vision narrows to a few degrees. The patient moves, twists, turns to different positions, trying to escape the intolerable. The world has shrunk down to a tiny space around his body, which is drenched and shaking. He is aware only of the cot on which he sits in the emergency room, itself rattling noisily with his shaking, and dimly of the people coming and going, trying to determine what has exploded his white count and made the intolerable pain. His body is emptying itself through all its fissures, but any concern for courtesy is forgotten along with the rest of the world. Eventually the drugs and intravenous therapy begin their stabilization and the world expands a little. Then it contracts again as diagnoses are considered and fear displaces some of the pain.

This time I am the patient, and in the intervening periods I can reflect on the experience of a collapsing world and our means of escape. In what must be a common experience, I feel trapped, incarcerated in a body that is not functioning properly; in particular, that has diminished the support I need to organize and communicate and be free—that is, to live. I know the world's sudden collapse is unreal, and I know that pain is simply my brain's response to a profusion of incoherent signals—a kind of overwhelming noise—from its "connected" organs, capturing its processor and masking intelligent signals. Might this understanding offer a way to freedom, for me or for any of us in our various kinds of pain?

I review what I know about pain. The pain-noise and the brain's response are essential to survival of our particular life-support system because they provide warning and trigger rapid response. An important part of our programming and control systems are devoted to this reception and response. That has been essential in the evolution of human organisms. The noisy pain-signal, however, is still only an electrochemical transmission, the pain peculiar to our wiring. Thus our various anesthetic and psychological remedies all intervene in the transmission, reception, or processing of the pain signal, and we shall doubtless get better at the intervention.

But forms of pain we cannot cure will likely survive our intervention, and we are hoping for more than analgesics. My pain is the hard-wired type, acute and overwhelming, but we all know other types. The despair of loneliness and inadequacy drove two of my closest friends to suicide. Fear shuts down our processors. Anger and envy narrow our apertures. Soured relationships distort receptions. Grief smothers all external stimuli. Pain comes in all flavors, and we all get strong tastes from time to time.

Those non-hard-wired pains are responses to signals too, in their case originating outside our "connected" domain, carried in by our six senses (augmented by increasingly extensive instruments.) These receptions, however, do not constitute our whole experience. Einstein considered "the happiest thought of my life" his realization that the concentration of energy and matter determined the geometry of space-time. This happy experience was "internally" generated, although it of course depended on his prior receipt of information generated by others.

Less happy experiences are also internally generated, as in the case of my two friends. Their despair proceeded from some combination of internal anxiety and insufficient or flawed communication with others and in themselves, perhaps intensified by genetic predisposition. We all have felt such moments, when hope is lost or grief or loneliness overwhelming. It is much like physical pain in the shrinking of the perceived world, the shortening of reach, the cutting off of aspiration.

* * *

ALL OUR EXPERIENCES have in common a dependence on communication—among neurons or among brains, within a close community of associates or between more distant persons not intentionally conversing, direct or mediated, amplified or attenuated by others. (Although we distinguish by habit between "external" and "internal" signals, their impact on experience is not necessarily sensitive to such a distinction.) In the unhappy experiences, something has gone wrong in the communication—perhaps what we call organic malfunctioning, or in more distant information transfers. In any case, the exchange of information or lack of it is determinative.

In communications parlance, all our pains can be considered forms of noise, interference, or distortion. Insistent, incoherent, likely in all channels, they mask the signals that support full participation in life as we have characterized it. But we have learned how to deal with such problems in communications: we can widen our apertures, tune our receivers, correlate receptions with known patterns, and integrate the output over many repetitions. This technology allows us to detect the tiniest signal from the greatest distances—to pull it up from enormously more powerful noise in which it lies embedded. [1]

What shall we listen for, then? What is the true signal embedded in the noise of pain and what is the specific "pattern" on which it is modulated, with which our receptions might be correlated? I have time here, and good reason, to ponder how one might attend more consistently to farther signals. In this I have help, not only from Einstein and Shannon, but from my two friends whose pain overwhelmed them. What signals would have made a difference to Winston and Carl, and might to me?

To be healing, I think they would testify, such signals must convey a true and fuller sense of participation in life. They should therefore include whatever communicates to our experience the presence of intelligence—evidences of order, of building, learning, healing, integration; evidences of freedom; discovery, the receipt of information; giving and receiving love, in which we extend our selves; and the perception, contemplation, and making of beauty (which we said, in the letter from Hard Scrabble Wash, is one of our great windows on life.) These evidences of intelligent life may be experienced ("heard") even in great pain, by focusing on them in both relation and contemplation.

All of experience is perception; but not all of reality. There is always room to increase perception, there exists always the opportunity to broaden experience—as any scientist and any artist will attest. As our own species evolves, its collective perception grows. So does our sharing of that perception, and the integration of our mutual experience. As a thought experiment, suppose we could extend such a sharing to include other intelligent "species" beyond our present knowledge, other life forms in our universe (or others.) I try to imagine what the perception and experience of a "whole people" (taken in the universal sense) would be. As such a whole is approached, their collective aperture would widen to look at more of reality, and the processing of this expanded information could be integrated more coherently. The perception of the whole people, and thus their experience, would approach ever more closely the discernment of all the real.

This integrated perception/experience would also tend to cancel out those parts of the members' perceptions that are *not* based in reality, as it accumulates shared observation. That is, the unreal perceptions would be more susceptible to "destructive interference" (to borrow a term from wave mechanics); they would have a higher probability of canceling each other, of disagreeing, than the perceptions of reality. Thus true perceptions gradually prevail.

* * *

EACH OF US, of course, suffers some combination of illusion, suggestion, and "noise" (to use the telecommunication term), varying over time that appears real at the time but is not:

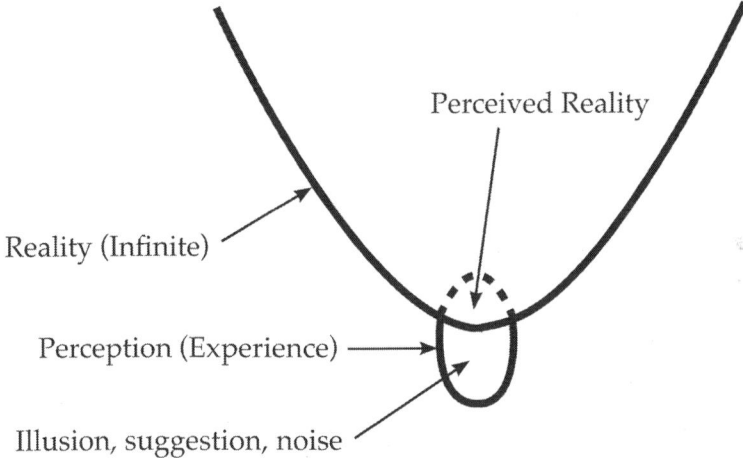

The part of each member's perception that lies within the reality circle is more likely to coincide with other members' perceptions than the part that lies outside, simply because it is true—it will pass every test, prevail in every event, return a correct answer to every question—whereas erroneous perceptions have higher probabilities of disagreement. Of course there are always commonly-held biases, time-honored errors, popular myths, widespread prejudices, and other shared illusions; but these have of necessity a lower probability of being shared indefinitely. So my personal circle of perception (experience) will grow and shrink, and will move back and forth across the reality boundary, depending on my observations and rationality and relationships—and my pain. So will yours, and all of ours. Even all of us together will not be always right. But the whole people's perception will progressively occupy larger regions of reality.

My whole experience, in fact, enters by the same portals that admitted my pain, the same that welcome beauty, that bring love, support discovery, allow communication and relation. My perception is my experience. Of course objective reality exists, but my experience of a given event is different from yours, which is different from others'. The whole of any concept or event could be perceived, experienced in its entirety, only by combining all views of it together.

Can I apply this concept to my own pain? Even in our limited world, humankind as a whole transcends the incapacities that have trapped me here and tormented my friends. Each individual suffers pains, but we do not all suffer the same pain at the same time. Each of us sleeps but we are not all unconscious simultaneously. So individual incapacities do not add up to disable the whole people, and we go on. When we cooperate in a joint endeavor, our individual efforts do add up coherently. Thus a difficult task, unyielding to individual effort or overwhelmed by stronger influences, gradually may be achieved through sustained cooperative work. Just as we open our receivers to admit the whole broadband signal, we combine the efforts of many people. Their individual pains are experienced randomly and therefore do not add coherently, but their cooperative, coherent effort persists with a cumulative effect.

We have the opportunity to work cooperatively toward any goal of mutual benefit, with a fair chance of reaching it. The more we cooperate, the less any individual's misfortunes or misbehavior impact the general welfare, which accrues to the eventual benefit of all the members. Self-interest is other-interest. Every board of directors, every committee, every family or community or association, is a model testing this general theory.

Of course each isolated individual in our experience suffers pains, incapacities, and death. These sufferings are starkly drawn on the grounds from which I write to you. None of us must look far to encounter them. Yet as we work together, we heal our pains as a people. Smallpox is gone. Our telescopes orbit earth and map our universe. The secrets of the human genome slowly yield to our research, with promise of more universal comfort. We can light our nights, warm our homes, teach our children, share our arts, sustain justice in most places most of the time. That we fail to do so consistently for all is a deficiency not of capacity but of character. Our grandchildren will see discoveries of which mankind has dreamed for millennia and some that even our dreams have not reached. As a people, we increasingly elude pain of all sorts.[*]

This being whole is of course what I miss. Those of us in pain (of whatever kind) feel broken and isolated, whereas the people as a whole retain the capacity to love, to benefit each other, to build and learn together, to comfort, to see beauty and make it, to live more fully. I search for a way to experience what the people can experience, to perceive what they can see, to be whole again through their wholeness.

In times of pain, whether physical or psychological, we sense a fragmentation, a distance from being whole. Pain, it occurs to me, is the experience of isolation. As I lie here, I test this theory on other kinds of pain. Surely grief is just that, and loneliness, and the physical and mental disassociation we call death. Other forms of despair, fear, physical as well as mental illness, strike me as experiences of being cut off or fragmented. This suggests that pain might be assuaged by connection, by association. The means of association available to intelligence generally would involve communication in all possible forms.

---

[*] I have worked sufficiently on our streets and in our institutions to be purged of any delusions in this matter. We have the capacity to destroy each other too, and everything within reach, and sometimes it seems we are trying to do it. The difference, of course, lies in the degree to which we are committed to truth and mutual benefit.

* * *

So it is not so much escape as connection that I seek. I will consider, then, the hypothesis of an integrative nature of intelligence. Might I be "in touch with" it or, if intelligence is pervasive and primordial, "part of" it? I imagine intelligence free of anthropomorphic mechanics. Nothing inherent therein ordains separation. Communication—the exchange of signals carrying information and instruction—is essential to it but has no inherent borders. Organizing, building, recognizing and making order with a purpose, and the freedom to do so, all grow in proportion to aggregation and collaboration. I consider also identity, what we make of the life-capacity. I have hypothesized that it is expressed in large measure through relation. Its commitments and transparencies, to the extent it finds meaning, are directed to truth and to each other. So the trap of isolation is not innate in what we recognize as life, in either its quantitative or its qualitative aspects. It is integration for which we hope in times of pain, and we must look to each other to find it.

Certain limits that we have imposed on ourselves and each other are artificial. The acute narrowing of experience that comes with pain is an extreme case of more general boundaries we accept daily, but which prove no more inherent as they are examined closely. The collapse of the world around me *is* unreal. This fact goes beyond refuting the episodic narrowing of conscious awareness. Our world is always narrower in awareness than in reality. The experience of pain—although it seems most organic and inescapable—is not qualitatively different from other signals reaching the brain.

This suggests that we look beyond the close signals of pain, whether acute pinching or daily enclosing, toward broader realities. Perhaps I can strive to experience somehow a wider truth— a universal truth, not just local facts—such that what I perceive may be projected upon immediate experience. In the micro world, such a "projection" of a quantum state upon an elemental particle is occasioned by making a measurement; in particular, by gaining information about it, before which it exists in a superposition of multiple states. Roger Penrose even speculates that the "non-algorithmic" functioning of human intelligence may involve the resolution of such superpositions. [2] While one should not expect quantum-mechanical phenomena to be exhibited consistently in our macro-world, the central role of information is shared.

* * *

MY FIRST INSTINCT was to find an escape from this experience, a way out of the enclosure that was pinching my existence. Upon better consideration, it was a connection I wished to find. But it is more. What I need to perceive, to experience, is an inclusion in the whole of which I am part. One's perception of such inclusion should make a difference. The extent to which I perceive my inclusion in the whole of intelligence—the extent to which I perceive the extension of my being toward its full expression in the life we share—should affect my experience of wholeness. In such perception I might project more of the whole people's experience, which ultimately approaches the convergence of life with truth, upon mine. In this I believe one may find healing of fragmentation.

Occasionally we do experience a closeness that seems to dissolve the conventional boundaries of self. Your mom and I have had such an experience, with each other and with you and your siblings, making the sharing of life a cogent reality for us. In all real elements, this association has enlarged my experience of life. With respect to both my identity and my vitality, to the qualitative as well as the quantitative aspects of my experience, to my being as well as my living, I am entirely different from whom I would be without you and your mom. This is a rational conclusion. Quantitatively we consider life as the ability and freedom to organize, build, perceive, and communicate, and I am truly more free and more capable in all such respects. Qualitatively I can relate better to all, commit more faithfully, offer better transparency. These constitute living and being in their real and lasting meanings. This is a stretch, a growth, an expansion, a reach. It is the direction of becoming, together.

Our ancestors, our friends and family, fought and died for our liberty. Yet an isolated identity is not ultimately defensible ground. Our visits to Met State and the South Bronx show how vulnerable we are from both inside and outside. Our brains, daily more understood, can be manipulated by electrical, chemical, or psychological interventions, either purposeful or accidental, and some day will be predisposed and even replicated by genetic programming. Even in sleeping or sickness or stress or fatigue, we must admit to the variability of our life-expression. Our commitments and transparencies, which are a part of who we are, vary also, voluntarily and involuntarily. Finally, Gould reminds us that the definition of an individual is itself imprecise. (See the letter from the Marmottan Museum.) Holding on to an absolute, isolated individuality is much like holding on to an irrational religion, defending a shrinking realm of mystery against scientific inquiry. We must find our freedom not in a fortress but in exploration, not in isolation but in integration.

The evolution of life as we know it is a story of connection, of growing complexity, of developing communication. The human form may seem a culmination of this process, because of where we are today, but it is part of a continuum. There is little doubt that the human family will continue to evolve, and its greatest potential lies in its capacity to build and organize together, to reach out to all of life, to enhance its freedom in communication. The phantom limits that pain imposes on us are dissolved in both our stretching and our becoming.

Each of us is made more free by relation to the other; each has an enhanced ability to perceive order, to build, to heal, to understand; each communicates more fully, not only with the other but beyond. Moreover, each of us strengthens her commitment and transparency to truth and the benefit of others through the experience of love, and the quality of relation. These are the capacities that constitute the experience and manifestation of intelligent life, so such a stretch is not merely emotional, nor is it mystical. I consider it a rational explication of the potential for enlarging "self," just as my earlier experience of acute pain shows the potential for its shrinking.

Downstairs in Obstetrics there are expressions of life coming into our view every hour. We can consider every strand of intelligence, in whatever form, part of a whole. It may be that we are surrounded, perhaps in dimensions of time and space not perceived, by expressions of life quite different from ours. In fact that seems more likely than a hypothesis claiming our own type as the only one evolved, or even possible. There is no reason to expect that every form of life should communicate using the media and modulations that humans have developed, so perhaps we shall not be able to reach them all, but this is not an excuse for not trying.

One's expression of life develops from communication and progresses in freedom. One does not come into the world or leave it with an isolated, constrained identity. Who one is proceeds from the choices one makes—one's commitments and extensions and transparencies. This varies. Identity is not fixed. But this variable, imperfect, vulnerable, indistinct person is included in the Whole Each Other.

The beneficial reach is toward greater integration, not inward. The truth of my "self" is found in its relation, and of my life in its sharing. To focus on the particular package that we observe at this immediate point in evolution is to miss the reality and the potential. A better focus would be on the integrative nature of intelligence. It is that integration that we wish to experience: the reality of a being and a living not stalled or contained.

We see so little in general, and our pains reduce our vision further. We are so incomplete. Our means of perception are limited to the six senses evolved in response to earth's peculiar demands.

Our media of communications are still just chemical, tactile, and electromagnetic (see the letter from MIT.) Our experience, and thus our psychological construct, limits our expectations. We live in a sample of reality, conditioned to its boundaries, not seeking the whole. Thus I believe we see only a part of ourselves and others, but can strive to perceive the unseen extent, the disregarded possibilities.

* * *

WELL, HERE IS the Rev. Natalie Wellrock, come to visit. I tell her I am not ready for last rites, which she finds less amusing than I do. Natalie shares my love of good writing, and is good at it herself. She led me to authors important to me, and I have shared a little of my early attempts with her. Now she is interested in how my rational faith is holding up in practice as well as in explication, and whether I can tolerate using the "G-word."

Indeed, I have avoided reference to God, because the word is so loaded, so evocative of widely-variant concepts and emotions. But Natalie uses a debater's trick on me: "Tell me about this God you don't believe in." Of course her point is that many have been turned away from religion by assuming it requires acceptance of a creed or mysticism or irrationality that is not necessarily demanded. Natalie's God, it turns out, is entirely consistent with my devotion to Truth and a Whole Each Other, and she suggests that I could use the G-word rationally.

I ponder this after she leaves. It is said that there are no atheists in foxholes, that in our extremes we all reach for any God that might be there. This is of little comfort, however, to those who have insisted until then on a rational faith. If we are to pray, it had better be a rational act with a rational object, or we shall not believe our own appeals.

Therefore we must seek our comfort, as we have our meaning, in truth. It will not do to seek a suspension of physical laws that have passed every test, or to turn away from new information or honest counsel, since we have subscribed to a rational faith that rests on a commitment to truth, unflinching, wherever it leads.

Such a faith, however, does support the stripping away of notions that are not grounded in truth, and the reaching for wider realities. We do not strive for miracles but for perception of our real inclusion in truth and life. So we may seek beyond limits that are merely habitual, boundaries to which we are merely conditioned. The containment of life in discrete anthropomorphic packages is an example. We accept this containment because we are long conditioned to it; but the capacity and freedom to organize and communicate knows no such bounds. This is a shared capacity and a freedom curiously expanded by joining. Even the life expressions with which we are most familiar work by collaboration, from the cellular level to the most complex organizations of human enterprise. We live, not just with, but in each other.

Of course we wish things to change. All the places from which I write to you threaten us in one way or another. At the Fitness Club we agonized over our lack of attractiveness, on the streets of the South Bronx over our violence toward each other, at Met State over the vulnerability of our very personalities. The "slings and arrows of outrageous fortune" afflict all of us, and a rational prayer does not seek to deflect them from oneself to others. Rather we seek to help each other, and we have recourse to each other; this has been the means of our progress, increasing the people's comfort over eons.

Praying to change the condition of this isolated person, this arbitrary aggregate, is of no use. It changes all the time, for and against our will. But it is not the whole story. We cannot change the past, which exists as "remembered information," or the present which passes instantaneously. We can, however, influence the future, which is not entirely determined by what has gone before, and we can stretch our reach and our experience toward more of the whole.

Our recourse is to each other, not only over generations of research and civilization, but in immediate crisis. We come down from the hill at Met State and away from Matt's crib and out of the Berkshire cemetery with no other answer to despair. Yet this is, I believe, the whole answer. We find our being and our living in each other, and by extension in the Whole Each Other.

A rational prayer could be an expression of trust (of faith) in the Whole Each Other of which we are a part, in whom we are included. It could be, at the same time, a discernment of Truth, prevailing in every event. It is true that we share life as we understand it and that the identity of each of us includes always an element of relation. It is true that intelligence need not be limited to the isolated physical structures in which we feel confined. In distress we can recognize and insist on such truths, and look hard at each other, to find there not a temporary containment but an expression of life, not an end but a becoming, not a detached individual but part of a Whole. This trust and this discernment constitute a rational prayer, simply to Truth and to the Whole Each Other.

At the start of a new day, at the end of the day, and in times of joy, an appropriate prayer would express gratitude for the prevailing of truth in every event and for the sustaining of life through our nights; shame and remorse for being unfaithful; and commitment to truth and the benefit of life in all circumstances. This helps to keep one's focus on the real meaning of life, internalizes the lessons learned, tests one's hypotheses against accumulating experience, strengthens one's identity, and prepares one for the opportunities and challenges that will arise. In times of pain and stress, whatever the source, a rational prayer would seek extension and inclusion of oneself in the Whole Each Other, reaching beyond the narrowed experience of confinement and loss.

Prayer, of course, should not be an event but part of being. One should strive to *be* thankful, not just to give thanks; to *be* remorseful for infidelity, not just to beg forgiveness; and to *be* committed and transparent and related and truthful.

Will this really make a difference, will it change things? I do not ask it to. What I want is to perceive what is already there. This appeal to truth, this rational form of prayer, derives its effectiveness from expanding one's own perception of reality. It does not seek to work a miracle or call down a supernatural intervention, but to support a larger discernment of what is real. Rational prayer is the listening for Truth.

I propose as a logical and a rational conclusion, therefore, that we are included in what I have called a Whole Each Other—not a heavenly, detached, supernatural Being, but the entirety of intelligence. One may or may not accept the hypothesis that such intelligence is pervasive, intrinsic, and primordial (as opposed to a local consequence of evolution), but, like the invisible "fields" that make up physical reality, we are in it and of it. (By "we" I mean the intelligence manifested by and among us.) I can't see the four fields or forces that hold our universe together either—gravitational, electromagnetic, weak and strong nuclear forces are not easily grasped—but I believe they are there and our world would fall apart without them. The experimental verification of intelligence and its existence as an intrinsic reality seem to me no less convincing.

* * *

WE HAVE CONSIDERED an expression of faith equivalent to an expression of love. Can we consider our love of a Whole Each Other to be reciprocated? Is there a rational equivalent to what the religious call the "love of God"? May I draw comfort from being "loved" by the Whole of which I am part? To be consistent, the love of such a Whole would be unconditional and would constitute the fullest expression of commitment, trust, and extension: the answer to this longing should lie in finding in the universe a perfection of these elements of love. Indeed I propose they are there: perfect and unconditional commitment is expressed in the prevailing of truth in every event, in the sustaining of life through our nights; perfect trust in the infinite potential of our becoming; and perfect extension in our inclusion in the Whole.

That inclusion is a fact. That we fail to discern it as love owes to our narrow definition. Love seems to us exclusive and personal; to the Whole it would be inclusive and universal. The gift of the Whole to each is the gift of truth and life, with all their vast potential. What each makes of that potential is of course the reciprocal gift we can make to the Whole. In such a gift, if faithful to truth and the benefit of all, we "become," in Whitehead's sense; we participate in the "creative advance of the world." In this we complete the consequent, derivative nature of the conceptual, primordial ideal of truth and life. The equivalent, reciprocal process completes our natures by perfecting relation.

In love, we commit and trust and extend our selves to another, which expands our identities. Such extension in love is achieved through truthful and beneficial (i.e., faithful) communication. In faith, it is through the equivalent process of "becoming"—of progressively more faithful expression of commitment, transparency, and relation to truth and the Whole Each Other. Integrating this concept of love and faith yields a full expression of identity: the integration of commitments into Commitment; of trust into Transparency; and of extensions (as we have said) into Relation.

Extension is an essence of love and of faith, just as are commitment and trust. In a close experience of love, my being expands—our beings intermingle, overlap, mutually influence each other. We "identify with" each other, much as the overlapping fields of microscopic particles develop a whole new form as their wavelengths stretch at very low temperatures. To extend to the Whole Each Other would require identifying with the perfection of truth and life.

How shall I seek such an ultimate extension, such an identification? Logically, it would require (a) truth in my behavior and my seeking of truth; (b) my use of the life-capacities—of learning, building, loving, healing, beauty—for the benefit of others; (c) my sharing of information, communicating freely; (d) my transparency to all these, and (e) my commitment to all these. This truthful way of living and being would integrate into the ultimate relation, with the Whole Each Other.

One's expression of the "relation" aspect of identity is the integration of extensions to others. Extending to the Whole Each Other would complete the integration, removing limits on one's being. In extension to the Whole I complete the Relation element of my identity, and reciprocally I participate in the consequent nature of the Whole. To perceive the "love of God," I must see that God is part of me, and I of Her. This growth of being would have infinitely more potential and be without the possibility of loss.

* * *

IN THE CONTEXT of such a universal love, one is drawn to consider the possibility of a continuity of individual consciousness. This is an ancient question but our encounters suggest some new perspectives upon it. The opposite of life is disorganization, disintegration, disorder, incarceration, non-communication. Such a death is easily inflicted in degrees by ourselves or others, and indeed we all suffer it daily, in greater or lesser extent. The fear of death, like pain, is a fear of separation.

Shall we remain hopeless of any continuity beyond the permanent silencing of one's brain cells? If we are to be rational, faithful to truth, we must look upon this silence without unfounded hope of resurrection. The stumbling block is our focus on the "individual." We hang on to this concept, not comforted by the continuity of the whole of life, not ultimately satisfied by the sustaining of shared intelligence. The experience of this particular brain, of course, is limited to its infinitesimal memory plane and the capacity of its inputs and storage. (Indeed, if we should find a way—and I expect you will—to preserve and selectively replace memory cells, thus enabling indefinite continuity of individual consciousness, I would not count it an unblemished success.)

Suppose we were to re-phrase the question. May I count on a continuity of my participation, my inclusion, in the Whole Each Other, and in particular on a continuity of that expression of my identity that we have called "relation"? I believe the answer to that question is yes.

It is relation, we have observed, that life and identity—the quantitative and qualitative sides of intelligence—have in common. Relation is supported by communication; it expresses both our living and our being. Relation expresses the direction of our "becoming," that stretching of our beings toward the Whole. It is the loss of relation—of connection, of communication, of mutual extension—that we really fear. And to the extent that our commitments and transparencies are directed to each other, it is in relation that we express our full being.

In fact it is necessary to our concept of intelligence as both conceptual and consequent that its *entirety* be preserved. It is not whole if any part was lost, and each of us is a part. Such a concept does not admit disintegration except by choice of isolation.

So I do not seek continuity of a static "individual" consciousness at all. I do seek continued and growing connection in life with others, and have stretched in that direction in love relationships (and in faith toward the Whole.) I do not seek survival; I seek becoming. I do not wish to perpetuate an incomplete identity but to complete my being in relation to the Whole.

If I try to hold on to a limited individual experience, if I tightly embrace an exclusive, defined, static identity, I die a little every day. Only in stretching toward the whole—in extension—is my being truly expressed. My whole nature is not static, not defined by "this muddy vesture of decay."

The logical direction of intelligence is toward integration. Intelligence necessarily accumulates information, which is not limited. Communication is essential to all forms of intelligence, which must exchange and process information in order to function (to organize and learn.) Thus intelligence naturally shares information in growing circles, and the logical direction of intelligence is toward integration.

The logical direction of becoming is toward the fullest expression of life, both quantitative and qualitative. Each of us is free to choose such direction. It is the stretching of our narrow sense of identity, a connecting to the Whole Each Other, a discernment of a wider reality, a rejection of isolation. It is the natural fulfillment of life—that great improbability made of freedom and ordering and communication, whose intrinsic consequence in truth must be to grow and join and share. Our support in pain comes from our sharing of this life, our extension beyond arbitrary boundaries, our inclusion in the Whole.

We can widen our apertures. Those who love and learn, who build and heal and see beauty, have made a local choice in this direction, and experience its potential. I can find no reason to consider that potential limited. The opportunity, when confronting the myriad pains that afflict us and the specter of death that haunts us, is not so much to change perception/experience as to *widen* it. We may listen with patience at the frequencies of life, and integrate what is continuous there—not lost, not destroyed, not silenced, not distorted, not overwhelmed, not cut off, but only masked by the louder and wider noise of pain. The signals of life—the messages of building together, of beauty, of love—are not susceptible to loss but are merely obscured. This to me is the "still, small voice" heard by

Elijah, the voice of Gabriel in the Prophet Muhammad's ear, the truth of Dharma, the music to which Shiva dances.

Notes

[1] See the letter from Radio Communications Corporation.

[2] Roger Penrose, "Where lies the physics of mind?" in *The Emperor's New Mind* (New York: Penguin Books, 1991), 405–449.

Harvard, Massachusetts

I HAVE RETURNED TO the old radio telescope on the hill in Harvard to ponder again its perspective on what Edward R. Harrison called the "masks of the universe." [1] Each time we peel off a worn veneer of fuzzy thinking and erroneous theories, a fresh universe appears to us, internally consistent and apparently valid. Yet in little time cracks appear, and wrinkles in the finish, evidencing something that doesn't fit our model after all, and the peeling begins again. Harrison suggests we shall never see the "real thing," that our models merely approach more closely but cannot capture all of reality.

Are we close to removing the final mask, or will our next model be found incomplete as well? Are there more "universes" than the one that evolved us, probably with their own time and space and possibly different physical laws? Will we find evidence of Kaku's ten dimensions? [2] Are there many other life forms, and will we ever be able to communicate with them? And where, in all of this, is our home?

Seeking the "real thing"—even if we cannot ever understand it fully—seems essential to our living together, to our hopes and commitments, to who we are and who we might be. Every people has its creation story, and all yearn for the transcendent. The understanding we develop of the universe, where we fit in it, what makes us up, whether we are alone, how we might communicate beyond our present limits, provides grounding for what we choose to devote ourselves to, in study and in relation. The world's religions, its civilization, and its course of research have had at their base an implicit model of the universe. Our future moral compass awaits direction from our evolving model.

In such an understanding we also place our hope for relief of our deepest pains. In the finding of a home in a rational universe our grief may be assuaged, our identities stretched, our longing for relation joined, our constraints freed, our pains relieved, our fears calmed, our iniquity pardoned.

* * *

WHAT CAN WE can say with confidence? That our particular universe—its age, forces, physical constants, and series of random events—is a universe (perhaps the universe) that could evolve *us*, and that even miniscule changes in its characteristics or history would make us impossible. One cannot escape a feeling of awe in contemplating the combination of circumstances that has made our particular life form possible, and its infinitesimal likelihood of developing spontaneously.

Paul Davies devotes a whole book to the fantastic "coincidences" we find in our universe, without which life as we know it could not exist. In it he provides a review of the critical constants and their relationships, deriving the extreme implausibility of their perfect values. [3] Barrow and Tipler offer an exhaustive treatise on the same subject, providing an encyclopedia of derivations and references testifying to our wildly peculiar universe. [4] The universe's expansion and distribution of matter, the precise values of physical constants, and the long string of improbable events that constitute evolution, must all be just right—I mean precisely right—to get to us. The tiniest change in any of them, and we would not be here.

Reasoning outward from the unlikely fact of our existence, a number of "anthropic principles" has been postulated to explain the improbability. Of course the universe must have those special characteristics that will sustain (our kind of) life, since we are here to see it; and many of those characteristics can even be predicted from thinking through the events that would be necessary to evolve us. This is the "Weak Anthropic Principle" and just a consequence of logical thinking. Some have gone on to suggest that the unlikely coincidences are "required" in order to evolve life with the intelligence to observe it (the "Strong Anthropic Principle"). Barrow and Tipler review these coincidences, and the history of various scientific and philosophical explanations, in their sweeping survey of the concepts that underlie the several anthropic principles.

Many persons do see in these fantastic coincidences a divine intention, a Design. Paul Davies is quoted as writing:

> The very fact that the universe *is* creative, and that the laws have permitted complex structures to emerge and develop to the point of consciousness—in other words, that the universe has organized its own self-awareness—is for me powerful evidence that there is "something going on" behind it all. The impression of design is overwhelming. [5]

One cannot disprove such a hypothesis but it is not necessary to the narrow purpose of explaining the coincidences. Of course the long string of perfect values and evolutionary events is extremely unlikely, but to establish a probable Intent one would have to show that, at each step along the way, the alternatives were *more* likely. If one throws a large number of dice many times and computes the probability of the cumulative result, taking into account not only the numbers but the scattering and orientation of the dice, it will be shown to be infinitesimally small, nearly impossible to repeat. The point is that all the other possible results are also extremely unlikely (unless there is a bias in favor of some); there are just a very large number of possibilities.

So we find ourselves in a very special universe that supports our being here, a near-impossibility statistically, but not necessarily less probable than a huge number of other outcomes. Given that there is something rather than nothing, every possible combination of physical laws, constants, expansions from the initial singularity, and evolutionary histories is wildly implausible, because the process is so long and complex. Yes, we are special; but that alone does not establish Intent. We experience only that "universe" that supports us, with an indefinite number of others possible but perhaps beyond our knowing. If a very large number of outcomes is possible, the occurrence of our particular estate is not so astounding; we just see ours because it is the only combination of variables that allows us to be here.

It may be that many universes spring into being and out again, with only a few evolving intelligent life. Under such a hypothesis, we live in one of the few. Or it may be that a large or infinite number of universes exists, in which case our presence is not even unlikely, given eternity to work with. (The probability of a complex result that was infinitesimal with one attempt approaches a certainty as the number of attempts approaches infinity. The precise outcome of my dice-throwing experiment, for example, will eventually be repeated in every detail, if an infinite number of throwings is performed.) Many universes, although not so comforting if we cannot reach them, seem a reasonable hypothesis, one consistent with recent advances in theoretical physics.

But there is a third way to reason, not necessarily inconsistent with either statistics or Design. The third explanation suggests that intelligent life can take many forms and may be supported by/included in a more universal and primordial intelligence—that intelligence or consciousness is the "real thing," rather than just a consequence of one very lucky evolution in one very special universe. This is the line of reasoning followed by Schrödinger and others—an independent reality of intelligence/consciousness, not limited to particular forms, not dependent on the local time of each universe, precedent and pervasive as well as consequent and localized.

* * *

ACTUALLY, THE THREE explanations are not all mutually exclusive. Either combination of intentional design or statistical evolution with intrinsic intelligence is logically allowable. Personally, I find Gould's argument of "contingency" more convincing than assumptions of either predestination or purposeful tinkering by an exogenous agent—but Gould's argument is not inconsistent with the Platonic reality of intelligence itself, any more than a Design argument would be. [6]

* * *

WHEN WE SAY "the universe," we often mean "all there is." Our "universe," however, with its own space and its own time that began in a singularity and may end in one, may not be all. In fact if we are uncomfortable with a Design answer to the question posed by Davies, Barrow & Tipler, and others, a many-possibilities hypothesis accommodates our being here with Gould's contingencies far more comfortably than a hypothesis of ours as the only universe possible. Suppose, then, that there are many, or even an infinite number of, "universes." Are there many different forms of intelligent life, depending on different local conditions?

Rational argument does not require localized consciousness, a carbon base for intelligence, or an Only Universe. We have great difficulty imagining alternatives to these familiar circumstances, but it is equally difficult to imagine the birth of time and space with our own universe, such that questions of where and when have no meaning "before" about thirteen and a half billion years ago. Our time and our space were born with our universe, in an initial singularity that cannot be grasped intuitively. In the relativistic and quantum-mechanical frames of reference that we must accept, however (given their experimental as well as theoretical evidence), unimaginable phenomena are no longer remote fantasies.

To an "outside" observer, the likelihood that intelligence would be found only in carbon-based forms, and that it would always be discrete and localized within such forms, would not be judged high, *a priori*. Limiting options to cellular forms imposes restrictions not necessarily bounding on all possible expressions of intelligence. To us, however, other forms seem alien and unwelcome to contemplate, because they are unfamiliar. This is a result of conditioning, not only to our own familiar forms but to the apparent inseparability of conscious life from conceivable support systems. "Life" is most often defined in terms of DNA replicability, but that is not what we listen for on Harvard's hill. The signal that we are not alone would not be a wave or a handshake or an attack, but evidence of an extra-terrestrial capacity and freedom to choose order, to organize, to communicate. Even our special universe does not require "biological" housing for such a capacity.

If an infinite number of universes is possible, then an infinite number of life forms is also possible (in fact, statistically necessary if the possibilities are played out); and even in our own universe there is surely room for many forms. We might consider, then, whether the evolution of intelligence itself would likely follow a common path in all these cases. While imagining all possible "life forms" may be unrewarding, if we consider what we mean by intelligence we may find a more tractable paradigm. A generalized evolutionary scheme of intelligence would proceed at two levels: the quantitative (called life) and the qualitative (called identity.) Their "becoming" would proceed in parallel—although not necessarily at the same pace. (When the qualitative lags the quantitative, which it surely has in our civilization, we run the danger of destroying each other.)

This process is based on growing connection. If one pictures many nodes, each with a finite chance of connecting to a neighbor, then the growth of complexity is inevitable. Such a picture looks like a familiar "cellular" process, but it could be any growth of networks among active nodes, whatever their form. The essential ingredients, whether in a biological cell or some unfamiliar analog, are (a) the availability of energy to sustain organization against the slope of entropy and (b) *communicative* capacity. The "nodes" of intelligence are neither distinct nor fixed, but vary in time and intensity (per Gould, the "legitimate individual" can be recognized at many levels). [7]

At a primitive level of organization, the progressive increase of order can be seen in the growth of any complex crystal. Although the spontaneous development of such a lattice is easy to see, it cannot proliferate indefinitely because its genesis is in cooling—the surrounding entropy must increase, and the process can be spontaneous only locally, not generally. A preference for wider and longer-term organization depends on a consistent bias in random processes—a recognition and choice of order. A spontaneous genesis of such a recognition and freedom are not so easy to see. Once started, however, intelligence likely confers a strong selective advantage in any evolutionary scenario—and it could "start" by accident. This is the process of evolution as we understand it, sustaining traits of advantage in feeding, fighting, and procreating. It is not limited to biological progress, however: the selective advantage is conferred generally by connection, cooperation, even at the statistical level.

The quantitative process depicted above is inevitable. To sustain a benefit among its members, however, its qualitative parallel must trend toward mutual commitment and relation as it builds communication and order. Although our own civilization presents an imperfect example, the quantitative and qualitative elements of growth are symbiotic. In mutual commitment and strengthening relation, the nodes of intelligence experience "extension" in each other, with a trust that opens apertures for increased communication. This permits greater organization and freedom (the capacity to affect probabilities), such that self-interest is equivalent to other-interest.

Then the question becomes why? Why is there something instead of nothing, and why should we find in that something the astonishing presence of freedom, organization, and communication—those improbable capacities that Prigogine and Stengers call the "far from equilibrium" condition? [8] What "calls" intelligence into fullness? As Stephen Hawking puts it, "Why does the universe bother to exist?" [See note 9 in the letter from Gloucester Cathedral.] Insight can again be found in Whitehead's model of the consequent "becoming" toward the conceptual. He speaks of a primordial, conceptual nature of what he calls "God," and a consequent, derivative nature. (I have used the term "Whole Each Other" to connote an equivalent idea.) Whitehead says all actual entities "become." This is consistent with statistical evolution within intrinsic intelligence.

In statistical terms, we can picture this as a "Bias toward Life." At the primitive level, there is no memory and really no "time"— the bias there simply makes continued aggregation more probable than disaggregation. At higher levels, primitive memory develops. "Freedom" evolves as memory develops, giving meaning to the capability to influence probabilities of "future" events. The Bias toward Life works through a preference for order and communication, and then through a preference for relation and commitment.

* * *

OUR OBSERVED UNIVERSE is very special to us. Other universes might well support other forms of intelligent life—indeed they probably do, given the success against all odds of our own. We would recognize such other forms not by their physical similarities to us but by their freedom and capacity to organize and communicate.

That is what is peculiar to intelligent life. That is what is special, is improbable. There—wherever intelligent life develops—is where "Nature tolerates improbabilities sustained against her downward slope," as we said at the start of our conversation.

Here on earth, carbon-based, DNA-replicating brains evolved to express intelligent life. Elsewhere, other expressions are likely. They are hard to imagine—but less so than imagining that our form is the only one. Another form of life cannot be "alien," as we understand intelligent life. The capacity and freedom that constitute life by definition support our communication, our recognition of beauty, our seeking of truth, our love and faith. Each is a function of life. These capacities and freedoms that constitute intelligent life—that allow us (and others not yet known) to find meaning and joy and worth—these capacities and freedoms could be found in any place or time or form in which intelligence is possible. And if one subscribes to the third explanation of our peculiar estate—that intelligence or consciousness is an intrinsic, precedent reality as well as an evolutionary reality—they would be found widely.

Our humanity, which has accomplished so much and destroyed so much, which has courageously reached for universal good and brutally suppressed aspiration, that humanity has been circumscribed by walls of its own building. We are of course thoroughly conditioned to our walls. More than accepting, we defend them fiercely. They have become our comfort. The isolation of individual brains, the importance of tribal jealousies and national patriotisms, the mating, nest-building, territorial, proprietary instincts that evolved by necessity, have carried us through natural selection and given us tradition, have been home to us.

* * *

HOME HAS MEANT for us, as we have wandered, a close-knit group, a shelter, and a refuge. Home is familiar, contextual, non-challenging, certain. It is a place for the tired to rest, the stressed to relax. Is it not time, however, to build more stately mansions? Holmes challenges the mature people to leave their low-vaulted past, to reach for a wider being in a more universal sharing. [9]

A primitive society forms exclusive, secret, and defensive groups; a mature civilization seeks an inclusive community. A primitive society hides in shelter from a feared environment, and exploits what it can; a mature civilization seeks to understand and preserve its environment. A primitive society takes refuge from others and erects defenses; a mature civilization seeks connection, common interests, and a way to build together.

Home is an island for the primitive, but for the mature, it is the universe. For intelligent life, the means of connection is communication. We have only begun to understand how to communicate with each other and with the Whole. I commend to you that search.

Once again I follow the antenna's gaze from Harvard's hill, and wonder at what it can see that we cannot. It occurs to me that I have known where it is looking, after all.

* * *

IT IS LOOKING home.

Notes

[1] Edward Harrison, *Masks of the Universe* (New York: Macmillan, 1985.)

[2] Michio Kaku, "Part II: Unification in Ten Dimensions" in *Hyperspace: A Scientific Odyssey Through Parallel Universes, Time Warps, and the 10th Dimension* (New York: Anchor Books [Bantam Doubleday Dell], 1995.)

[3] Paul Davies, *The Accidental Universe* (Cambridge University Press, 1982.)

[4] John D. Barrow and Frank J. Tipler, *The Anthropic Cosmological Principle* (Oxford University Press, 1988.)

[5] Paul Davies quoted by Chet Raymo in "God of Ignorance, or of Knowledge?" *The Boston Globe*, 1 May 1995.

[6] Stephen Jay Gould, *Wonderful Life: The Burgess Shale and the Nature of History* (New York: W. W. Norton & Company, 1989.) The entire book is relevant to this point, but see especially pages 288–291.

[7] Stephen Jay Gould, "A Humongous Fungus Among Us," *Natural History Magazine*, July 1992.

[8] Ilya Prigogine and Isabellet Stengers, *Order Out of Chaos* (New York: Bantam Books, 1984.)

[9] Holmes, Oliver Wendell, "The Chambered Nautilus."

> Build thee more stately mansions, O my soul,
> As the swift seasons roll!
> Leave thy low-vaulted past!

Glossary

Adams, Dr.

A psychiatrist working for the Commonwealth of Massachusetts, in charge of the Region containing the prison, state hospital, and school for the retarded from which three of these letters are written.

Barrow, John D.

Lecturer, Astronomy Centre, University of Sussex

Bronowski, Jacob

Mathematician, inventor, physicist, literary editor, Visiting Professor of History at MIT, Deputy Director of Salk Institute, educator, host of television series on "The Ascent of Man" in 1974, prolific author/lecturer on the history and philosophy of science and its correlation with the arts and civilization.

Buber, Martin (1878-1965)

Jewish scholar, author, interpreter of Chassidism, translator, and Israeli mediator. Goethe Award (Univ. of Hamburg), Erasmus Award (Amsterdam), Faculty of University of Frankfort and Hebrew University, international lecturer. Elucidator of the radical potential of "relation," as summarized in his best-known work *Ich und Du (I and Thou)*.

Capra, Fritjof

Physics researcher and lecturer, author and lecturer on similarities between modern physics and Eastern mysticism.

Davies, Paul C. W.

Physicist, author, broadcaster, public lecturer, Professor of Theoretical Physics, University of Newcastle upon Tyne; Chair of Mathematical Physics and Professor of Natural Philosophy, University of Adelaide (Australia). Author of over 100 research papers on cosmology, gravitation, quantum field theory, and related topics, and over twenty books on scientific and philosophical subjects. Winner of 1995 Templeton Prize for progress in religion.

Eliot, Abigail Adams (1892-1992)

Founder of the Eliot-Pearson School, now a department of Tufts University. Prominent in early childhood education and community mental health. First cousin of T. S. Eliot. Friend of the author.

Faith

Commitment to, trust in, and extension of one's self toward, the object of one's faith. (If a person, such faith would be called love.) A rational faith would include truth among its objects.

Gould, Stephen Jay

Researcher, teacher (Harvard University and elsewhere), and author in the sciences of biology and geology and related historical and philosophical subjects. Winner of numerous awards including a MacArthur Fellowship and the National Book Award.

Harrison, Edward

Professor of Physics and Astronomy, University of Massachusetts.

Hawking, Stephen

Physicist, author and lecturer, prominent researcher into the basic laws of the universe. Lucasian Professor of Mathematics, Cambridge University. Recipient of a dozen honorary degrees and numerous awards. Fellow of The Royal Society, member of the U.S. National Academy of Sciences.

Heisenberg, Werner

Physicist and Nobel laureate, one of the pioneers of quantum mechanics. Professor of Theoretical Physics at the Universities of Leipzig, Berlin, Gottingen, and Munich. Director of the Max Planck Institute of Physics. Developer of matrix mechanics and formulator of the Uncertainty Principle that bears his name.

Hofstadter, Douglas R.

Professor of computer science and cognitive science, Indiana University. Winner of the Pulitzer Prize (for *Gödel, Escher, Bach*).

Holmes, Oliver Wendell (Sr.)

> Physician and poet. Professor of Anatomy and Physiology, Harvard Medical School.

Hoyle, Fred

> Astronomer, physicist and prolific author of science and science fiction. Lecturer in astronomy, Cambridge University.

Identity

> One's expression of commitment, transparency, and relation.

Information

> Information is the reduction of uncertainty, as quantified by Shannon's theory. Quantitatively, information is proportional to the uncertainty in any message: it peaks at a point where all possible messages are equally probable, and drops off as any message (or symbol) becomes more likely than others. Once received, therefore, the information content of any message goes to zero (for that recipient). I have called this (at the moment of its receipt) "actual information." Received information, however, adds (as long as it is remembered) to one's store of understanding and to the context shared for future transmissions. I have called this "remembered information." The remaining uncertainty I have called "latent information."

Kaku, Michio

> Physicist, author, public lecturer and broadcaster. Professor of Theoretical Physics, City College of New York. Author of over seventy scientific articles and nine books (as of 2001), including the national best-seller and award winner *Hyperspace*.

Lederman, Leon

> Physicist and Nobel laureate. Discoverer of the muon neutrino and other basic particles. Faculty of Columbia, University of Chicago, and Illinois Institute of Technology. Director of the Nevis Laboratory at Columbia and the Fermi National Accelerator Laboratory.

Life

> Intelligent life would be recognized by its capacity to organize, to bring order out of (or recognize order in) random/noisy/

chaotic processes, combined with the freedom to choose such order and the proclivity to do so; and further combined with the ability to communicate. A freedom and ability to communicate and to bring about and sustain order constitute a far-from-equilibrium condition, in which may be expressed the capacities to build, to learn, to heal, to see or make beauty, and to love. Such capacities and freedom would not necessarily be limited to anthropomorphic or even organic agents, nor to our evolutionary prerequisites, nor to our means of sensing and communicating.

The "meaning" of such life would be found in seeking truth and the benefit of others, and in our "becoming" together—in supporting, advancing, and participating in that direction of growth and evolution that most fully expresses the capacities of intelligent life and its potential for integration toward a whole people.

Manchester, William

Historian

Mario

A prison inmate and parolee from whom I learned more than he learned from me.

Markham, Edwin

Poet

Marsalis, Wynton

Musician and teacher

Matt

The young son of a close friend and colleague, Matt's presence among us was cut unfairly short by a disease that your children will heal. Seeking a response to such unfairness sustains this writing.

Mogul

A nickname given by the author, as an expectant parent, to the child for whom he waits, from whom he seems to hear a very long question, to whom he writes this long response. Mogul also represents the children of us all, whether parent or not, the children of the life we share.

Mostel, Zero

    Actor and humorist.

Mueller, Robert Kirk

    Businessman and author, director of many companies including one of the author's, Chairman of Arthur D. Little.

Pagels, Heinz R.

    Physicist and science writer. Associate Professor of Theoretical Physics, Rockefeller University. President of the New York Academy of Sciences and the International League for Human Rights, Fellow of the New York Institute of the Humanities at NYU.

Penrose, Roger

    Mathematician, physicist, author. Rouse Ball Professor of Mathematics, Oxford University. Discoverer of geometrical abstractions in higher dimensions and of many solutions to complex mathematical and physical problems.

Pollard, Mary and Oliver

    Bereaved parents in colonial America.

Pope, Alexander

    English poet, translator, and satirist.

Prigogine, Ilya

    Winner of the Nobel Prize for his work on "dissipative structures" in the thermodynamics of non-equilibrium systems. Researcher and writer on complex inter-disciplinary problems, in Brussels and at the Ilya Prigogine Center for Statistical Mechanics and Thermodynamics, University of Texas.

Question

    The question raised by the Mogul, by our looking anew upon life, is a long question: how shall we give to our children, and to their children, a rational faith, grounded in truth and not in mystery, a faith that will last, and a way to live, a way to meaning, worth, and comfort, a way not reliably found in religion or science alone but true to both; how shall we increase our understanding of those fundamental concepts that light our way and our becoming; how shall we find a meaning of life, a purpose to our evolution, a basis of worth attainable by all, a reliable moral compass and an ultimate justice; how shall we offer a comfort in the myriad pains we suffer, physical and emotional and spiritual pains, the pains of living and dying; and how shall we explore a primordial and timeless universe, a home of intelligence, a confluence of life and truth, and our inclusion therein?

Rogers, Carl

American psychologist of the humanist school, proponent of the innate drive toward self-actualization.

Sagan, Carl

Astrophysicist, Harvard and Cornell. Researcher into extraterrestrial intelligence. Prominent teacher and broadcaster.

Schrödinger, Erwin

Physicist and Nobel laureate. Pioneer of quantum mechanics, developer of the "wave mechanics" formulation.

Shannon, Claude E.

Engineer and theorist, developer of the mathematical formulation of information theory.

Slope, Nature's downward

The second law of thermodynamics observes the universal tendency toward decay, disorganization, and equilibrium imposed by probability in all events. Few processes are entirely reversible; i.e., it takes more energy to run things backward than was expended going forward, because some of the original energy went into a chaotic form called heat. (That heat, divided by the temperature at which it was released, is called "entropy," which is the measure of disorganization.) This "downward slope" is reversed only by the continuous application of purposeful energy, as in building or organizing or pumping things up against their natural falling. The most highly-organized, far-from-equilibrium, low-entropy things are least probable—for example, the sustaining of intelligent life.

Stengers, Isabelle

Philosopher, chemist, and historian of science, collaborator with Ilya Prigogine.

Tallis, Thomas

Musician and composer, prominent in the re-birth of English music in the sixteenth century, under four monarchs.

Thomas, Lewis

Physician and author.

Tipler, Frank J.

Professor of Mathematics and Physics, Tulane University.

Truth

That which prevails in every event, passes every test, and returns a perfect answer to every question; the perfection of intelligence, the integration of all information.

Whitehead, Alfred North

Mathematician, co-author (with Bertrand Russell) of *Principia Mathematica*, author and philosopher.

Index

A

ablative living, 119–120, 122
abstract thinking, 19*n*, 146
acceptance, 29, 36–37, 81, 93, 95, 104, 112
accomplishment, 65–66
achievement, 68
alien life, 81–82
Allen, Jonathan, 157
analogy, 87, 197, 201, 214
Anasazi, 146
Anthropic Cosmological Principle, 17, 242, 250*n*, 274–275
appearance, 65–66, 68
Arlington Street Church (Boston, Mass.), 13–14, 178, 211
arrows of time, 125, 131–132, 135, 221–223
art, 15, 32–33, 53, 77, 130, 139–140, 143, 145–149, 151–153, 209, 243
artificial intelligence, 19, 28, 80, 85–86, 165, 167–168, 247
arts, 35, 79, 87, 152–154, 163, 201

B

Bach, Johann Sebastian, 14–15, 22, 27
balance, 32–36, 152, 162, 170
baptism, 20, 22
Barker, Francis, 54–55
Barrow, John D., 17, 274–275, 277
beauty, 32, 79, 126–127, 130, 139145–154, 213–214, 243–244, 256,
becoming, 80–81, 119, 163, 180, 186, 209–210, 212–214, 229–230, 249, 262, 266, 268, 270–271, 278, 280
Beethoven, Ludwig van, 32, 69, 136, 153, 243

behavior, 59, 109, 131*n*, 133–134, 136, 209, 231–233, 269
beliefs, 74, 178
Bell Labs, 9, 12
Berkshires (Mass.), 193, 197, 266
Bethe, Hans, 157
bias, 37, 275, 279–280
big bang, 9, 73
Blake, William, 153
Bolton (Mass.), 194, 197
Boltzmann's Constant, 250*n*
Bose, S. N., 84
Bose-Einstein condensate, 84
Boston (Mass.), 13, 63, 117, 209–213
Boston Industrial Mission, 157
Brave New World (Huxley), 171
Brief History of Time, A (Hawking), 131
Bronowski, Jacob, 33–34, 37, 79–80, 86–87, 163, 166, 197, 201
Bronx 2000, 227–228
Buber, Martin, 71, 74, 92–93, 97, 108, 110–112, 183–184, 196–197, 199
building, 19, 86, 110–111, 114, 120, 126–129, 135, 158–160, 163, 170, 196, 198, 202, 212, 229–232, 241, 256, 259–262, 269, 271, 282

C

Cambridge (England), 165
capacities of life, 94, 280–281
Capra, Fritjof, 20–21
caring, 104, 113–114, 122, 199
Carroll, Lewis, 218
cause and effect, 132–133, 189
central order, 185–186, 197, 200, 202, 204, 245
Chandrasekhar, Subrahmanyan, 150
Channing, William Ellery, 13

chaos, 15, 18, 87, 151–152, 196, 203, 239

choice, 28–29, 36, 78, 108, 131, 133–134, 161, 167, 196, 198, 208, 234, 241

Christianity, 127, 130, 178

churches, 20, 119, 179, 182, 233
    Catholic, 127–128, 182
    history, 14, 175–178

civilization, 95, 119, 130, 157, 162–163, 166, 179, 183, 193, 198, 231, 274, 282

coherence, 17–18, 97, 241, 246

coincidence, 274–275

Coleridge, Samuel Taylor, 79, 202

collaboration, 114, 134, 229, 260, 265

comfort, 87–88, 103, 121, 178, 191, 195, 229, 259, 264–265, 281

commitment, 66–70, 73–74, 88, 93–95, 101–104, 108–113, 120, 122, 134, 139, 151, 158–163, 168–169, 180–182, 184, 188, 191, 195, 198–200, 202, 227–231, 233, 235, 246, 248, 260, 262–263, 267–270, 279

communication, 18–19, 27–37, 38n, 39, 111-114, 129, 131–132, 135, 146–148, 158, 160–163, 166–167, 208–209, 214, 253–256, 258–265, compassion, 110, 114, 219, 233

Concord (Mass.), 55, 194

conditioning, 278, 281

connections, 73–74, 77, 79–80, 82, 86–87, 99, 102, 104, 120, 140, 142, 146, 148, 150, 163, 165, 168–169, 190–191, 197, 200, 203, 234, 246, 248, 259–260, 262, 270, 279, 282

consciousness, 19, 78, 80, 84–87, 109, 130, 133, 140, 161, 164, 166–169, 185, 189–190, 200–202, 219, 233, 260, 269–270, 275–278, 281

context, 29–32, 34, 36–37, 73, 78, 80, 93–95, 100, 104, 112, 129, 147, 152–154, 161–162, 187, 199, 208, 213

convergence, 245

cooperation, 80, 82, 183, 187, 196, 246, 258–259

Coronado, Francisco Vásquez de, 127

correlation, 129–130, 148, 170

cosmological time, 131, 183

courage, 54–55, 100, 122, 171, 197, 228, 231

creativity, 80, 86–87, 136, 152

creed, 232

Crick, Francis, 165–166

D

Dancing Wu Li Masters (Zukav), 21

Davies, Paul, 169, 196, 274–275, 277

death, 109, 122, 176, 193–195, 227, 258–259, 269, 271

design, 275–277

despair, 4, 87, 193, 254, 259, 266

Dillard, Annie, 126, 140–141, 148–149

discovery, 256, 258

disorder, 125, 130–133, 167, 198, 203, 221, 269

Distant Mirror, A (Tuchman), 176

diversity, 27, 83, 168–169

Dorchester (Mass.), 117–120

E

Eddy, Mary Baker, 182

education, 231

Einstein, Albert, 53, 69, 84, 109–112, 197–198, 200, 219–220, 254–255

electromagnetic waves, 11, 83–84

electron, 73, 84, 218–219, 220n

elegance, 35, 38n, 150–151, 197, 245

Eliot, Abigail Adams, 14, 71–72, 117–122, 197, 199–201, 211

Eliot, T. S., 117

Emerson Hospital (Concord, Mass.), 1–4, 27, 212, 249

Emperor's New Mind, The (Penrose), 19

energy, 97, 131, 158, 203, 237–241, 250*n*, 254, 279

engagement with others, 102–103, 119

engineers, 149, 157–159, 203

entropy, 28, 30, 34–37, 38*n*–40*n*, 64, 127, 132, 158–159, 165, 189, 203–204, 221, 237–243, 246, 248*n*, 279

equilibrium, 1, 81, 131, 133, 189, 222, 239, 241

evil, 228, 232–235

evolution, 95, 114, 119, 134–136, 159–160, 164–165, 186, 196, 242–244, 249, 262–263, 267, 273–281

extension, 71–72, 93–95, 103–104, 110–114, 122, 162, 165, 180, 182, 185, 191, 231, 261, 263, 266–271, 280

extraterrestrial intelligence, 8–10, 17–18, 34, 85, 164–165, 183, 185, 198–199, 278

F

failure, 25–26, 46–47, 60

faith, 22, 66–67, 74, 83, 93–94, 109, 119, 122, 127, 130, 150, 158, 161, 168–169, 171, 176, 178–182, 184, 190–191, 199, 213, 234, 262, 265–269, 281

fear, 87–88, 97, 103, 108, 164, 204, 228, 254, 259, 270, 274

Feynman, Richard, 223

fidelity, 29, 35–37

Fitness Club, 63–74, 88, 212, 265

forgiveness, 16, 70, 112, 227, 234, 267

free will, 133–135, 245

freedom, 18–19, 36–37, 49, 52–55, 78–79, 86, 133–134, 167–168, 195–196, 262

Frost, Robert, 7, 33, 93, 142, 193, 212

future, 207–208, 222, 224, 265

G

genetics, 59–60, 103–104, 135, 165, 254, 259, 262

Gloucester Cathedral, 175–176, 178, 182, 185, 212

God, 72, 101, 133, 164, 179, 185, 194, 199, 247–249, 264, 267, 269, 280

Gödel, Kurt, 202, 248

goodness, 121, 228–229, 232–235, 281

Gould, Stephen Jay, 77–78, 81, 160, 188, 196, 262, 277, 279

grace, 92, 234

Green Mountains, 91–92, 212

grief, 4, 10, 70, 193–195, 254–255, 259, 274

H

Hale, Edward Everett, 118

happiness, 170

Hard Scrabble Wash (Arizona), 145, 152, 212, 256

harmony, 152, 246

Harrison, Edward, 211, 273

Harvard (Mass.), 7, 27, 128, 130, 151, 273, 278, 282

Hawikkuh (Zuni land), 125–128, 130, 212

Hawking, Stephen, 131–133, 187–188, 204, 280

healing, 3–4, 19, 67, 69, 81, 86, 109–111, 127, 158, 196, 202, 231, 233–234, 256, 262, 269, 271

health care, 1–4

Heisenberg, Werner, 133, 169, 185–186, 189–190, 197, 200, 202, 220, 245, 246*n*, 248

heroes, 69–70

heroin, 43

Hodge, Frederick Hodge, 128

Hofstadter, Douglas, 30–31, 80, 85, 248
Hofstadter's Law, 100
Holmes, Oliver Wendell, 282
holy, 88, 168, 213
home, 281–282
honesty, 119
hope, 4, 146, 178, 180, 199, 249, 254
Hopi, 146
Hoyle, Fred, 85
humanity, 281
humor, 32, 99–102, 104, 163
Huxley, Aldous, 171
hypotheses, 149, 164, 188–189, 202–204, 211–214, 228, 244–245, 263, 266–267, 275–277

I

identity, 10, 20, 29, 43, 47–49, 59, 64–66, 68–73, 77, 81, 83–84, 88, 93, 95, 103–104, 110–114, 118–119, 135, 142, 147, 161–163, 185, 195–200, 204, 229–230, 246, 249, 260–263, 266, 269–270, 274, 278
ignorance, 53, 177–178, 241n
illness, 81–82
illusion, 257
individuality, 72, 77–79, 109, 121, 196–197, 232, 262, 269–270
influence, 73–74, 84, 217–220, 224
information, 18, 27–29, 32, 132, 135–136, 142–143, 147, 151, 159, 166, 170, 189, 197, 202, 208–210, 213, 218, 220–224, 234, 241, 243, 245–246, 248–249, 254, 256, 260–261, 265, 269, 271
information theory, 28, 30, 35, 38n, 93, 95, 112, 166, 208, 241n
information transfer, 31, 35–36, 113, 147, 161–162, 255
insights, 80, 87
instincts, 134–135

institutions, 57–61, 111
integration, 110–111, 166, 170, 197–198, 249, 256, 260, 262–263, 268, 271
intelligence, 1, 18–19, 78–80, 85–87, 135–136, 151, 160, 164–165, 167, 169, 190, 200, 202, 204, 208, 220–221, 242–249, 256, 259–261, 263, 266–267, 270–271, 275–280
intelligent life, 8–10, 17–19, 27–28, 31, 34–35, 37, 67, 85–86, 94–95, 99, 104, 108–109, 118, 127–128, 131, 134, 149–151, 159–160, 162–163, 167, 183, 185, 189–190, 195–196, 198–199, 202, 231, 241, 244–245, 256, 263, 276–277, 280–282
intention, 180, 275–277
isolation, 73, 110, 171, 197, 199–200, 232, 246, 258–259, 262–263, 265–266, 270–271

J

Johnson, George, 218
Jon, Uncle, 52, 55
Jones, Mal, 8
justice, 179, 232–233
Justice Research Institute, 232

K

Kaku, Michio, 136, 183, 203, 273
Kechipauan, 128
Kennedy, John F., 69
Kepler, Johannes, 201
kinetic sense, 104n, 203

L

laboratories, 25–28, 37, 84, 136
latent information, 221–222
learning, 19, 67, 80–81, 86, 109–111, 127, 135, 158, 196, 202, 249, 256, 259, 269, 271
Lederman, Leon, 21, 217–218

life, 1, 9, 15–19, 29, 65, 69, 71, 74, 78, 86–87, 94, 97, 104, 109–110, 118–122, 126, 130–131, 133, 135, 139, 142–143, 150–152, 154, 157–158, 160–165, 169, 171, 185, 187–188, 195–196, 198, 208–209, 224, 229–230, 234, 244–247, 249, 253, 255–256, 259, 261, 265–266, 268, 270–271, 273–281. *See also* intelligent life
Lincoln, Abraham, 69–70
Lincoln Memorial, 53
"Location of Meaning, The" (Hofstadter), 31
loneliness, 254–255, 259
love, 92–97, 101–105, 112–114, 171, 179–180, 199–200, 267–270
Luria, Salvador, 157

M
machines, 136
magic, 87
Manchester, William, 177
Manhattan (N.Y.), 139–140
Mario, 41–43, 46–47
Markham, Edwin, 69–70
Marsalis, Wynton, 162, 166, 169–170
Massachusetts Institute of Technology, 157–164, 212, 264
mathematical concepts, 184, 186, 202, 204, 211, 223, 242–243, 245
Matt, 4, 26, 41, 249, 266
matter, 84, 204, 254
meaning, 67, 74, 88, 146, 161, 169, 172, 178, 244, 260, 264
media, 161–163, 190, 204, 263–264
medieval world, 175–177
memories, 80, 132
memory, 135–136, 140, 207–209, 221–224, 234, 244, 265, 270, 280
mental health, 118–119
mental hospitals, 58, 107–111, 228
mental illness, 107–111, 259

message transmission, 29–31
metaphor, 32–34, 86–87, 152, 201, 214
Metropolitan State Hospital (Mass.), 107–114, 212, 219, 249, 262, 265–266
Middletown Prison, 45–49, 60, 108, 212
mind, 4, 26, 79–80, 87, 130, 150, 166, 182, 190, 201–202, 242
Mogul, 1–2, 4, 10, 22, 26, 37, 72, 122
Monet, Claude, 77, 81, 140–142, 153–154
Monhegan Island (Maine), 99–105, 212
morality, 19, 104, 118, 178, 182, 186, 198, 233, 274
Mostel, Zero, 99–100, 102
Mozart, Wolfgang Amadeus, 242
Mueller, Robert Kirk, 217, 225*n*
Musée Marmottan (Paris), 77, 79, 141, 212, 262
Museum of Fine Arts (Boston), 209–210
music, 9, 11*n*–12*n*, 14–15, 26, 32–33, 136, 146, 152, 162, 177, 242–243
mystery, 88, 126–127, 143, 221, 231, 248, 262
mysticism, 148–149, 161, 190, 200, 264

N
NASA, 8
Native Americans, 102, 104, 125–128, 145–146
Natural History (Gould), 77–78
natural selection, 242, 251*n*, 281
nature, 9, 125–127, 148–150, 164, 179, 185, 195, 201, 214, 237, 249, 281
Navajo, 146, 151
Newfern School, 57–61, 108, 212
Newton, Isaac, 133

non-algorithmic consciousness, 80, 85–86
Norsemen, 102, 104
Northeast Utilities, 237–238
Northfield Mountain (Mass.), 237–238, 241, 249

O
Oak Ridge Observatory (Harvard, Mass.), 7–8, 10, 27, 128, 130, 151, 198, 212, 273, 278, 282
observation, 45, 96, 129, 135, 143, 151–152, 180, 184, 187, 189, 247, 256–257
"On Rainbows" (Sayre), 16
order, 15, 17–19, 32, 67, 72, 78–79, 81, 87, 94, 108–109, 126–127, 129–131, 131n, 133, 146, 150–152, 158–159, 162–163, 167, 169, 185–187, 189–190, 195–196, 198, 200, 203, 208, 214, 229, 231, 241, 243, 245–246, 256, 260, 262, 271, 278–280
order, central. *See* central order
organization, 17, 19, 67, 85–87, 109, 131, 131n, 135, 158–161, 163, 184, 196, 198, 209, 229, 233, 241, 244, 247, 253, 260–262, 265, 278–280
"Origin of the Universe, The" (Hawking), 187–188
other-interest, 79, 200, 228, 230, 258, 280

P
Pagels, Heinz, 109–110, 132, 135, 219, 248n
pain, 4, 83, 87–88, 92–93, 100–101, 111, 113–114, 121, 171, 176–177, 224, 228–230, 253–267, 269, 271, 274
particles, 83–86, 97, 134–135, 143, 185, 203, 211, 217–220, 223–224, 240, 260–261, 268

pattern recognition, 80, 128–130, 146, 158, 166, 255
Penrose, Roger, 19, 39n, 80, 85–86, 184, 242, 244, 261
Pentagon, 51–52
Penzias, Arno, 9
perception, 113, 122, 129, 131, 149, 152, 164, 167, 197, 222, 256–259, 261–263, 265, 267, 271
perpetual motion machines, 240
petroglyphs, 145–147
philosophical thought, 69, 79, 108, 110, 161, 165, 188, 213, 243, 275
photon, 219, 220n
physics, 20–22, 133, 135, 143, 151, 165–166, 187, 189, 203, 276
Platonism, 86, 94, 184, 242–245, 247, 277
poetry, 31–34, 93, 117, 142, 146–147, 152, 201
Pollard, Mary, 194–195, 197, 199
Pollard, Oliver, 194–195, 197, 199
Pollard Family, 194–195
Pope, Alexander, 54, 79–80
power, 52
prayer, 146, 179, 228, 264–267
predictability, 133–134, 166
Prigogine, Ilya, 131, 133, 241, 280
"Principle of Process" (Whitehead), 210–211
prison, 41–42, 45–49, 58–59, 107, 109–110, 112, 147, 219, 228, 233
prisoners, 45–49, 60–61
probability, 1, 38n, 40n, 78, 134, 197, 239–240, 242, 275–276, 280–281
psychology, 95–96, 110, 131–132, 222
purpose, 17–18, 160, 163, 178, 246, 260

Q
qualitative living, 229–231, 246, 260–262, 271, 278–279

quantitative living, 229–231, 260–
261, 271, 278–279
quantum mechanics, 86, 133–134,
165, 167, 202–203, 218–219,
223–224, 240, 241*n*, 260–261, 278
question, 10

R

Radio Communications Co. (RCC),
25–28
radio noise, 9, 11–12
radio telescope, 7–10, 18, 128, 166,
273
random events, 207, 246, 279
rational faith, vii, 22, 94, 159, 168,
171, 180–184, 190–191, 193, 213,
232, 234, 249, 264–269
rationality, 19, 22, 27–29, 67, 72–73,
83, 87–88, 94, 103, 119, 122,
127–128, 133, 142, 148–149, 151,
158–160, 168–169, 179–183,
185–186, 190–191, 200, 211, 213,
228, 232–233, 247, 249, 261, 269,
274, 277
real world, 209, 211–212
reality, 77, 81, 83, 87, 99, 104, 113,
121, 136, 143, 149, 151, 164, 168,
171, 178, 181, 184–187, 189, 196,
202, 212, 223, 228, 243–245,
256–258, 264–265, 267, 271,
273–274, 281
recidivism, 41–42, 46
recognition, 85–86, 162, 260
reductionism, 135–136, 142
redundancy, 32–33, 170
rehabilitation, 41–42, 60–61
relation, 54, 60, 69, 71–73, 86, 88, 95–
97, 104, 108, 110–113, 121–122,
134, 146, 152, 162, 185, 196–200,
214, 219, 228–231, 233, 235,
246–247, 254, 258, 260, 262–263,
266–270, 279–280
relativity, 77, 165, 202–203, 243

religious thought, 20, 69, 72, 87, 102,
104, 119, 146, 151, 169, 171, 179,
182, 184–188, 191, 214, 232, 249,
262, 264, 267, 272–275
remembered information, 221, 224,
234, 265. *See also* memory
resonance, 33–34, 83, 86, 140, 143,
146–147
Rogers, Carl, 95–96
Rouen Cathedral, 140

S

Sagan, Carl, 8–9
Sandburg, Carl, 101
sandstone, 126–127
Schrödinger, Erwin, 165–166, 169,
190, 197, 200–201, 276
Science and Human Values
(Bronowski), 33–34
scientific thought, 69, 79–80, 87, 148,
150–151, 154, 158, 163, 165–166,
169–170, 178, 181–182, 187–188,
197, 201, 203–204, 213–214, 223,
230, 249, 256, 262, 275
scientists, 28–29, 150, 158–159
Search for Intelligent Life (SETI),
8–10
seeing, 139–141, 154, 282
self-extension, 110–111, 120, 261,
263
self-image, 47–48, 65–66, 68, 81–83,
104, 108–110, 114, 183
self-interest, 27, 29, 43, 68, 79,
161–163, 186, 200, 228, 230–231,
233, 258, 280
separation, 110, 200, 260, 269
SETI. *See* Search for Intelligent Life
(SETI)
Seven Golden Cities of Cibola, 127
Shaftsbury (Vermont), 91–93, 103
Shakespeare, William, 12*n*, 33–34
Shannon, Claude E., 28, 33, 35,
38*n*–39*n*, 93, 101, 132, 162, 166,

170, 197, 199, 203, 208, 221, 248, 255

shared experience, 83, 95, 100, 129, 147, 151, 153–154, 162–163, 167, 170–171, 178, 187, 196–197, 200, 256–258

shared identity, 20, 22, 27, 83, 86, 93, 95, 103–104, 109, 112, 114, 118, 135, 147, 200

sharing, 119, 135, 143, 159, 161–163, 181, 190, 200, 204, 213, 230–231, 256, 263, 269–271, 282

signals, 28, 83, 88, 97, 166, 204, 253–256, 260, 278

slope, 1, 97, 125–127, 131, 185, 249, 279, 281

society, 79, 282

South Bronx (N.Y.), 227–235, 249, 262, 265

space, 165, 211, 219, 223, 254, 263, 277

space program, 119–120

Spielberg, Steven, 8

Stengers, Isabelle, 131, 133, 241, 280

stretching, 72, 110, 130, 142–143, 197, 204, 262–263, 265, 270–271

string theory, 73

striving, 87–88

success, 65–66, 151, 187

Sullivan, Rev. Leon, 228

surprise, 31–33

symbiosis, 82, 187, 279

symbols, 30–31, 33, 36, 94, 146–147

symmetry, 129, 151–152, 214, 243

T

Tallis, Thomas, 177, 183

Tao of Physics, The (Capra), 20–21

teaching, 96, 158

Theresa, Mother, 235

thermodynamics, 28, 38n, 131–132, 203–204, 222, 239–240, 241n, 245, 248n

Thomas, Lewis, 2, 81–82

Thomson, J. J., 73

thought experiments, 84

Three Scientists and Their Gods (Wright), 28

Tilden Pond (Maine), 25, 207, 212

Tillich, Paul, 108, 249

time, 125–126, 131–132, 135, 165, 207, 211, 219–223, 247, 254, 263, 277, 280

Tipler, Frank J., 17, 274–275, 277

traditions, 104

transcendence, 178, 184–185, 188

transient experience, 207

transparency, 68–74, 88, 95, 110, 113, 120–122, 162, 185, 196, 198–200, 213, 219, 229–230, 233–235, 243, 246, 260, 262–263, 267–270, 273

Trinity College (Dublin), 165

trust, 70, 74, 83, 93–94, 103–104, 108, 111–113, 159, 180, 182, 191, 199–200, 202, 214, 228, 233, 266, 268, 280

truth, 22, 27, 32, 36–37, 43, 53–54, 66–69, 73–74, 77, 79, 82–83, 86–87, 94, 113, 118–119, 121–122, 136, 143, 148–149, 151, 154, 158–161, 168, 170, 178–179, 181–182, 184, 187–191, 202, 208, 214, 223, 228, 233–235, 243–245, 247–249, 257, 259–261, 264–269, 271, 281

Tuchman, Barbara, 176

U

uncertainty, 28–32, 36, 38n, 55, 84, 86, 112–113, 133, 162, 202, 207, 220–221, 241

Unitarians, 13–14

unity, 10, 79–80, 87, 150, 201, 203

universe, 1, 8–9, 18–19, 73, 77–78, 81, 84–86, 88, 92, 96–97, 99, 109, 125, 127, 129, 131–132, 135, 152,

159, 164, 168–169, 171, 176, 183,
185, 187–190, 199, 208, 219, 231,
240–241, 243, 248*n*, 259, 267,
273–278, 280, 282

V

values, 79, 109, 186, 274–275
veracity, 29, 36–37, 95, 104, 112
Vermont, 141–142
Vonnegut, Kurt, Jr., 153–154
vulnerability, 25, 55, 93, 101, 113

W

Wald, George, 157
walls, 53–54, 71, 79, 107–108, 120,
246, 281
Walpole Correctional Institution, 209
war, 52
Washington (D.C.), 51–53
Watson, James, 165–166
wave mechanics, 165
waves, 83–84
Wellrock, Rev. Natalie, 20, 22, 264
What is Life? (Schrödinger), 165
white noise, 30
Whitehead, Alfred North, 210–211,
229, 244, 247, 249, 268, 280
Whole Each Other, 136, 183–184,
190–191, 232, 263–264, 266–271,
280, 282
wholeness, 259, 261, 265–266,
268–271, 282
Wilson, Robert, 9
winning, 52
World Lit Only by Fire, A
(Manchester), 177
world view, 74
Wright, Robert, 28
writing, 140, 147, 152, 212, 264

Z

Zukav, Gary, 21
Zuni, 125–128, 130, 146
Zuni River, 125, 145